Jessica Burgner

Robot Assisted Laser Osteotomy

Robot Assisted Laser Osteotomy

by
Jessica Burgner

Dissertation, Karlsruher Institut für Technologie
Fakultät für Informatik,
Tag der mündlichen Prüfung: 28.01.2010

Impressum

Karlsruher Institut für Technologie (KIT)
KIT Scientific Publishing
Straße am Forum 2
D-76131 Karlsruhe
www.uvka.de

KIT – Universität des Landes Baden-Württemberg und nationales
Forschungszentrum in der Helmholtz-Gemeinschaft

KIT Scientific Publishing 2010
Print on Demand

ISBN 978-3-86644-497-3

Robot Assisted Laser Osteotomy

zur Erlangung des akademischen Grades eines

Doktors der Ingenieurwissenschaften

der Fakultät für Informatik
des Karlsruher Instituts für Technologie (KIT)

genehmigte

Dissertation

von

Jessica Burgner

aus Wuppertal

Tag der mündlichen Prüfung: 28. Januar 2010
Erster Gutachter: Prof. Dr.-Ing. Heinz Wörn
Zweiter Gutachter: Prof. Paolo Fiorini, PhD

Meinen Eltern
in Liebe und Dankbarkeit

Zusammenfassung

Mit dieser Dissertation wurde das weltweit erste Robotersystem entwickelt, welches eine Osteotomie mittels Laser ermöglicht. Die Osteotomie – das Durchtrennen von Knochen für Umstellungen, Versorgung mit Implantaten oder zum Freilegen des Situs – ist in beliebigen Geometrien und Schnittwinkeln mit konventionellen chirurgischen Instrumenten nicht realisierbar. Das Entfernen von Hartgewebe mit Hilfe von Lasern ermöglicht einen kontaktfreien und präzisen Abtrag in beliebigen Schnittformen unter Abwesenheit von einwirkenden Kräften und metallischem Abrieb. Weiterhin sind mittels Laserablation deutlich geringere Schnittbreiten in der Dimension des Laserstrahldruchmessers erzielbar.

Bisherige Forschungsarbeiten haben die prinzipielle Eignung des Lasers als Werkzeug für das Schneiden von Knochen gezeigt. Dabei wurden rein statische Experimentalaufbauten verwendet und die Ausrichtung von Laser und Knochen zueinander manuell durchgeführt. Der sehr feine Knochenabtrag und die damit erzielbare hohe Präzision, sowie die Rahmenbedingungen des Schneidprozesses, erreichen die Grenzen der manuell umsetzbaren Genauigkeit der Chirurgen. Deshalb erfordert die Laser-Osteotomie die Unterstützung durch Methoden der computer- und roboter-assistierten Chirurgie. Ein robotisches Assistenzsystem welches präoperativ geplante Schnitte am Patienten mit einem Laser als Schneidwerkzeug umsetzt, existierte bislang nicht.

Die schnelle und präzise Ablenkung des Laserstrahls durch einen Strahlscanner und dessen Positionierungen durch ein Robotersystem sind aus Applikationen zum Schneiden und Schweißen von Werkstoffen aus der Industrie bekannt. Die große Herausforderung ist es, den Laserschnitt auch nach mehrmaligem Umpositionieren des Strahlscanners durch den Roboter exakt fortzuführen. Erfahrungen aus der industriellen Anwendung von Lasern zum roboterunterstützten Bearbeiten von Werkstücken sind allerdings nicht ohne weiteres auf den medizinischen Kontext übertragbar. Der Patient als ungenau beschriebenes *Werkstück* mit seiner jeweilig individuellen Anatomie und Pathologie erfordert letztlich eine *Einzelstückfertigung*. Das in industriellen Anwendungen weit verbreitete einteachen von festen Bewegungsabläufen kommt deshalb nicht in Frage. Planungsmethoden für das Schneiden von Hartgewebe mittels Laser sind bisher nicht existent. Während in industriellen Anwendungen die Erhöhung des Durchsatzes sowie der Prozessgeschwindigkeit im Vordergrund stehen, liegt das Augenmerk in chirurgischen Interventionen auf der exakten Umsetzung. Weiterhin bedarf die Integration eines robotischen Systems in den Operationssaal einer genauen Betrachtung aus Sicht des Risikomanagements.

Die wissenschaftlichen Herausforderungen der vorliegenden Dissertation zum Thema „Roboter-assistierte Laser-Osteotomie" lauten wie folgt:

- Wie lässt sich das Schneiden von Knochen durch Laser mit robotischen Methoden realisieren?

- Welche Genauigkeiten sind mit einem System für die roboter-assistierte Laser-Osteotomie gefordert und erzielbar?

- Wie kann die Ablation von Hartgewebe präoperativ geplant werden?

- Welche Parameter sind ausschlaggebend um den Ablationprozess zu kontrollieren und zu optimieren?

- Welche chirurgischen Interventionen profitieren von der roboter-assistierten Laser-Osteotomie?

Im Rahmen der Dissertation wurde ein prototypisches Gesamtsystem für die roboter-assistierte Laser-Osteotomie entwickelt. Auf Grundlage der Methoden aus der computer- und roboter-assistierten Chirurgie, wurde ein Gesamt-Workflow für die roboter-assistierte Laser-Osteotomie konzipiert und die erforderlichen Prozessschritte entwickelt und untersucht.

Neben der Systementwicklung konzentriert sich die Arbeit auf die präoperative Planung von Knochenschnitten. Ein Schwerpunkt liegt dabei auf dem Transfer geometrisch definierter Schnitttrajektorien in ein Ablationsmuster. Der Laserablationsprozess ist durch die Verwendung eines kurzgepulsten CO_2 Lasers diskret. Das heißt, der Knochenschnitt entsteht aus der Konkatenation von einzelnen Laserpulsen, die jeweils kleine Volumina des Knochens herauslösen. An dieser Stelle spielt die Modellierung des Ablationsprozesses und seiner beeinflussenden Parameter eine ausschlaggebende Rolle für die Güte der Planung. Im Rahmen der Dissertation wurden die entscheidenden Parameter der Laserknochenablation unter Verwendung eines Strahlscanners untersucht und modelliert.

Im Anschluss an den Transfer geometrisch definierter Schnitte in ein Ablationsmuster muss eine geeignete Lösung für die Lage des Strahlscanners bezüglich des Patienten erfolgen. Dazu wurde eine Optimierungsmethode entwickelt, welche die minimale Anzahl von Strahlscanner Positionen automatisch ermittelt. Eine Simulationsumgebung rundet die präoperative Planung ab und ermöglicht vorab die Visualisierung der roboter-assistierten Umsetzung des Plans. Die Planung findet unabhängig von der verwendeten Roboterkinematik statt, wodurch Modularität erreicht wird. In der Simulation kann die Ausführung dann mit einem speziellen Roboter hinsichtlich Erreichbarkeiten und Machbarkeit evaluiert werden. Im Rahmen der Dissertation wurden dazu zwei serielle Roboter untersucht, ein industrieller Reinraumroboter und ein Leichtbauroboter.

Für die präzise Ausführung vorab definierter Schnitte ist die Kalibrierung und Registrierung des Gesamtsystems, insbesondere des Endeffektors in Form des Strahlscanners, von essenzieller Bedeutung. Im Rahmen der Dissertation wurden dazu Methoden entwickelt und evaluiert. Das Gesamtsystem wurde zur Evaluierung von Schnitten an tierischen Knochenpräparaten experimentell erprobt. Die erzielte Gesamtgenauigkeit von 0.4 mm zeigt die Eignung der entwickelten Methoden. Eine Risikoanalyse und -bewertung des Experimentalsystems wurde begleitend durchgeführt und beeinflusste Entscheidungen im Entwicklungsprozess hinsichtlich der späteren Einsetzbarkeit im Operationssaal. In Kooperation mit HNO-Chirurgen wurde weiterhin eine medizinische Applikationen definiert und in ersten Experimenten die Eignung des Systems evaluiert. Des Weiteren eröffnet die Laserablation als Instrument zur Bearbeitung von Knochen neue chirurgische Applikationen, beispielsweise die mikrochirurgische Knochenentfernung oder die individuelle Erstellung von Implantatbetten.

Im Zuge der Dissertation wurden weltweit erstmalig intraoperativ einsetzbare Methoden für die roboter-assistierte Laser-Osteotomie entwickelt und untersucht, welche die präoperative Planung komplexer Schnitttrajektorien erlauben und diese präzise umsetzen. Somit wird den Chirurgen ein neues Werkzeug bereitgestellt, mit dessen Hilfe zukünftig völlig neue Osteotomietechniken umsetzbar werden.

Preface

This doctoral thesis was accomplished during my work as researcher in the Medical Group (MeGI) at the Institute for Process Control and Robotics at the Karlsruhe Institute of Technology (KIT), the former Universität Karlsruhe (TH). This work was funded by the European Comission in the scope of the specific targeted research project *Accurate Robot Assistant* within the sixth framework programme for three years.

Danksagung

Eine wissenschaftliche Arbeit ist nur selten die alleinige Leistung eines Einzelnen. Vielmehr sind neben dem wissenschaftlichen Rahmen jene Menschen entscheidend, denen man als Doktorand auf seinem Weg begegnet. Da sind die Einen, die aktiv unterstützen und die Anderen, die ohne es zu merken einen wertvollen Beitrag leisten; vielleicht nur durch ein Wort oder ein aufmunterndes Lächeln. Im Folgenden möchte ich einigen der Menschen danken, die auf die eine oder andere Weise zum Gelingen meiner Promotion beigetragen haben.

Für seine Unterstützung und die Möglichkeit an seinem Institut forschen und lehren zu können, danke ich meinem Doktorvater Prof. Heinz Wörn. Besonders freue ich mich, dass Prof. Paolo Fiorini das Korreferat meiner Dissertation übernommen hat. Ihm gebührt besonderer Dank für die inhaltlichen Diskussionen, bei denen er mir neue Sichtweisen eröffnete. Dem Leiter der Medizinrobotikgruppe am IPR, Jörg Raczkowsky, bin ich ebenfalls dankbar. Er gab mir die Gelegenheit neben meiner wissenschaftlichen Arbeit, auch meine Fähigkeiten im Projektmanagement, der Projektbeantragung und der Forschungsorganisation auszubilden.

Bedanken möchte ich mich bei Prof. Georg Eggers für zahlreiche Gespräche in denen er mir immer wieder voller Begeisterung seine spannenden Visionen für die Mund-Kiefer-Gesichtschirurgie vermittelte. Außerdem danke ich Priv.-Doz. Thomas Klenzner für die Anwendungsszenarien meiner Methoden in der HNO-Chirurgie. Weiterhin möchte ich mich bei Prof. Rüdiger Dillmann und Prof. Tanja Schulz bedanken, die mir in Gesprächen wertvolle Hinweise gegeben und mir als Prüfer eine angenehme Disputation bereitet haben.

Meinen Studenten die im Rahmen ihrer hilfswissenschaftlichen Tätigkeit, Studien- oder Diplomarbeiten zur Realisierung meiner Ideen beigetragen haben, danke ich sehr herzlich. Meiko Müller, Yaokun Zhang, Christoph Platzek, Kerstin Wimmer, Johannes Schick, Farzaneh Milani und Lydia Pintscher: ohne sie wäre mein System nicht in der Formvollendung in der kurzen Zeit entstanden!

Allen Kollegen am IPR, die mich bei meiner Arbeit unterstützt haben, möchte ich danken. Insbesondere gilt dies für Matthias Riechmann.

Unsere regelmäßigen Besprechungen zum Stand der Dinge im Projekt *Diss* und sein Korrekturlesen haben mir geholfen das Dokument abzuschließen. Meinem Zimmerkollegen Markus Mehrwald danke ich für die angenehme Atmosphäre, Darts Spiele und aufmunternde Postkarten. Ein besonderer Dank gebührt Prof. Ramon Estaña für die vielen Gespräche und Fünf-Minuten-Pausen, die mir unvergessen bleiben. Bei Björn Hein, Andreas Schmid, Hayo Knoop und Dirk Göger bedanke ich mich für tolle gemeinsame Konferenzreisen. Weiterhin bedanke ich mich bei Oliver Weede und Gavin Kane für die konstruktive Zusammenarbeit. Vielen Dank auch an Rene Matthias für das Korrekturlesen. Bei folgenden MeGI Alumni möchte ich mich für die schöne gemeinsame Zeit bedanken: Oliver Schorr, Michael Aschke, Helge Peters, Christoph Schönfelder und Annika Wörner. Suei Jen Chen danke ich u.a. für einen unvergesslichen Tag mit dem Kuka LBR.

Meinen Freunden und Kollegen an der Universität Verona danke ich für die schönen Forschungsaufenthalte. Neben den fachlichen Diskussionen war es ihre italienische Gastfreundschaft die mich in manch kritischer Phase des Schreibens bestärkt hat. Stellvertretend für alle möchte ich hier ganz besonders Debora Botturi und Francesca Pizzorni Ferrarese danken.

Dem Team des Institutssekretariats danke ich herzlich für die Unterstützung in allen organisatorischen Fragen. Allen voran möchte ich mich bei Elke Franzke bedanken, die in allen Belangen immer ein offenes Ohr und Ratschläge für mich hatte. Bei Hartmut Regner bedanke ich mich für die Geduld und Unterstützung bei meinen zahlreichen Werkstattaufträgen.

Bei meinen Freunden bedanke ich mich für ihr Verständnis und die motivierenden bzw. ablenkenden Worte. Ihr seid die Besten!

Meinen Eltern Dietlind und Artur Burgner gilt mein größter Dank, den ich mit Worten nicht auszudrücken vermag. Ich bewundere sie für ihren Mut und ihre Stärke, mich meinen Weg gehen zu lassen und mich dabei in jeder Entscheidung zu unterstützen. Bei meiner Schwester Melanie möchte ich mich dafür bedanken, dass sie mir oft auf ihre Weise zeigt, dass so manches weniger wichtig ist, als es mir erscheint. Ich habe sie nicht vergessen.

Meinem Partner, Lüder A. Kahrs, danke ich besonders für sein kritisches Korrekturlesen, die vielen verständnisvollen und bestärkenden Worte, sowie den Jungs. Für immer und einen Tag!

Karlsruhe, Januar 2010   *Jessica Burgner*

Contents

German Summary   i

Preface   v

Acknowledgement   vii

1   Introduction   1
   1.1   Scientific Challenges   3
   1.2   Structure   3

2   Laser Material Processing   5
   2.1   Laser Fundamentals   6
   2.2   Light Matter Interaction Mechanisms   9
      2.2.1   Reflection and Refraction   9
      2.2.2   Scattering and Absorption   10
   2.3   Absorption Processes in Matter Caused by Laser Radiation   11
   2.4   Industrial Laser Applications   13
      2.4.1   Laser Cutting   14
      2.4.2   Laser Welding   16
      2.4.3   Laser Drilling   20
      2.4.4   Laser Micro Structuring   21
      2.4.5   Conclusion   22
   2.5   Interaction of Laser Radiation with Biological Tissue   23
      2.5.1   Photochemical Interaction   24
      2.5.2   Thermal Interaction   25
      2.5.3   Photoablation   27
      2.5.4   Plasma-Induced Ablation   27
      2.5.5   Photodisruption   27
   2.6   Pulsed Laser Ablation   27
      2.6.1   Biological Tissue Properties   28
      2.6.2   Energy Response   29
      2.6.3   Thermo-Mechanical Response   31
      2.6.4   Thermodynamics   31

2.6.5 Ablation Plume Dynamics for Short Pulses 31
2.6.6 Ablation Models 32
2.7 Medical Applications Using Laser 35
2.7.1 Hard Tissue Ablation Using Laser 36
2.7.2 Thermal Effects 38
2.7.3 Healing 40
2.8 Conclusion 40

3 Computer and Robot Assisted Surgery   **43**
3.1 Computer Assisted Surgery 43
3.1.1 Medical Imaging 44
3.1.2 Image Processing and Representation 44
3.1.3 Planning 45
3.1.4 Registration 46
3.1.5 Navigated Surgery 48
3.2 Surgical Robotics 50
3.2.1 Milestones of Surgical Robotics 50
3.2.2 Interaction Categories 52
3.2.3 Workflow of Autonomous Robot Assisted Interventions 53
3.2.4 Mechanical Design 53
3.2.5 Bone Treating Robotic Applications 54
3.2.6 Current Developments 55
3.3 Computer and Robot Assisted Medical Laser Applications . 56
3.3.1 Conceptual Work 56
3.3.2 Urology 57
3.3.3 Orthopedics 58
3.3.4 Oral and Cranio-Maxillofacial Surgery 58
3.3.5 ENT Surgery 61
3.3.6 Ophthalmic Surgery 61
3.3.7 Laser Osteotomy 61
3.3.8 Preliminary Work at the Own Institute 63
3.4 Conclusion and Open Questions 65

4 Modeling the Discrete Ablation Process   **69**
4.1 Fundamentals 70
4.2 Preliminary Work 71
4.3 Specific Experimentation Methods and Parameters 72
4.3.1 Measurements of Removed Bone 73
4.3.2 Data Analysis for Parametric Description of Ablation
Craters 74
4.3.3 Energy per Pulse 76

      4.3.4 Usage of a Scan Head 79
      4.3.5 Optical Path, Spot Size and Shape 82
  4.4 Influencing Parameters 83
      4.4.1 Angle 84
      4.4.2 Distance 86
      4.4.3 Depth 88
      4.4.4 Catalyzing Fluids 90
      4.4.5 Thermal Dispersion 90
  4.5 Cutting by Pulse Concatenation 92
      4.5.1 Theoretical Considerations 92
      4.5.2 Strategies 92
      4.5.3 Experimental Evaluation 95
  4.6 Conclusion 101

5 Planning Robot Assisted Laser Ablation   **103**
  5.1 Cutting Geometry Planning 104
      5.1.1 Cutting Path 104
      5.1.2 Cutting Trajectory 106
  5.2 Transfer into Ablation Sequence 108
      5.2.1 Constraints 108
      5.2.2 Transfer of a Cutting Geometry into an Ablation
            Pattern 108
  5.3 Optimal Execution Planning 110
      5.3.1 Problem Formulation 110
      5.3.2 Two-dimensional Solution 112
      5.3.3 Three-dimensional Solution 114
  5.4 Projection into Scan Plane 117
  5.5 Conclusion 119

6 System Realization for Robot Assisted Laser Osteotomy   **121**
  6.1 System Fundamentals 122
  6.2 Hardware Components 123
      6.2.1 Laser System 123
      6.2.2 Articulated Mirror Arm 125
      6.2.3 Galvanometric Scan Head 127
      6.2.4 Controlling Laser and Scan Head 127
      6.2.5 Robot 128
      6.2.6 Experimental Setup 130
  6.3 Calibration and Registration 131
      6.3.1 Calibration and Registration between Scan Head and
            Scan Field 132

Contents

6.3.2 Registration between the Tool Center Point and the
      Scan Head Coordinate System 139
6.3.3 Patient Registration 140
6.4 Workflow 141
6.5 Software Modules 143
      6.5.1 Planning Module 143
      6.5.2 Simulation Module 146
      6.5.3 Execution Module 148
6.6 Conclusion 149

7 Experimental Feasibility Studies   **151**
7.1 Marking Experiments 152
      7.1.1 Preparation 152
      7.1.2 Planning 152
      7.1.3 Execution 152
      7.1.4 Evaluation 155
      7.1.5 Results 155
7.2 Cutting Experiments 156
      7.2.1 Femoral Cow Bone 156
      7.2.2 Cadaver Skull 158
      7.2.3 Evaluation 160
      7.2.4 Results 161
7.3 Medical Application 163
      7.3.1 Cochleostomy 163
      7.3.2 Preparation 163
      7.3.3 Planning 164
      7.3.4 Setup Trial 166
7.4 Conclusion 167

8 Discussion and Outlook   **169**
8.1 Discussion 171
8.2 Outlook 175

A Risk Analysis   **179**
A.1 Step 1: Intended Use and Identification of Medical Device
      Characteristics that Could Impact Safety 179
      A.1.1 Intended Use 179
      A.1.2 Reasonably Foreseeable Misuse 180
      A.1.3 Identification of Safety-Related Characteristics . . . 180
A.2 Step 2: Identification of Hazards 180
A.3 Step 3: Risk Evaluation and Risk Control 188
      A.3.1 Risk Evaluation 188

A.3.2 Risk Control 189

B   Mirror Arm Specification   **191**

C   Robot Specifications   **195**
C.1 Stäubli RX90B CR 195
C.2 KUKA LWR 196

Bibliography   **197**

1

Introduction

Long before human could read or write, they opened the skull of fellow men. Likewise extraordinary: The patients usually recovered from this procedure. Thousands of discovered specimen indicate that trephination, i.e. drilling or cutting out rings or squares, was mostly conducted after head injuries at these times. The *surgeons* applied sharped flints or obsidian fastened to a wooden handle. Later these instruments were replaced by copper or bronze blades. Nowadays mainly mechanical tools are applied to separate bony tissue, i.e. burrs and saws. However, mechanical processing of human hard tissue with these instruments constricts the choice for the incision execution. For example the osteotomy as a surgical procedure to cut bone in order to change alignment, provide an implant or to access the operation site is not realizable in arbitrary geometries and cutting angles with conventional surgical tools. Since surgical instruments are usually handheld, the precision of the cut is mainly dependent on the skills of the surgeon.

Known from industrial chipping applications, the major drawback of mechanical cutting instruments is the development of high temperatures at the cutting edges during processing. The rotational or oscillation movement of instruments causes the bone to heaten up. This can cause necrosis or carbonization leading to regeneration delay or at worst irreparable thermal damage. Furthermore, the cutting is directly dependent on the applied force. By-products of cutting, e.g. metal abrasion, have an additional impact on the bone regeneration process.

In order to overcome these disadvantages of mechanical cutting instruments new technologies were proposed over the last years. Beside the water-jet method, cutting with light was introduced shortly after the invention of the laser in the 1960s. Material processing using laser became a widely used method for industrial cutting applications. This is due to several advantages, which are unique for this energy source: non-contact processing, improved quality, higher productivity and automation worthiness.

Recent publications revealed, that short-pulsed carbon dioxide (CO_2) laser ablation is suitable to ablate bony and cartilage tissue. In comparison to conventional cutting methods, laser ablation facilitates cutting widths which are one dimension smaller (down to the laser spot size). Furthermore the cutting geometry is no more restricted. Geometrically arbitrary and complex cutting shapes are achievable. Laser ablation also allows processing of bone in regions, which are hardly accessible with conventional mechanical instruments. Hence, fundamentally new operation techniques become feasible.

However, every tool leaves its own fingerprints in the tissue. In case of short-pulsed laser ablation the bone removal is driven by the thermo-mechanical process taking place when irradiating the tissue with laser pulses. This means that the energy absorbed by the mineral and water components of the bone leads to a fast rise of the temperature inside the bone. Increasing pressure promotes the explosive removal of tiny bone fragments which are carried away with the vapor. In order to avoid carbonization of the bone by too high temperatures, a fast scanning of the focused laser beam over the tissue is proposed. The laser beam multi-passes the bone and thereby the incision is performed layer by layer. Histological evaluations revealed that short-pulsed CO_2 laser ablation does not have any negative impact on the bone regeneration. However, up to now only the principal feasibility of short-pulsed CO_2 laser ablation to cut bony and cartilage tissue was shown.

Narrow incisions with only a few hundred microns necessitate precise guidance and application of the focused laser beam. To achieve such accuracy manually is beyond the human capabilities. Hence, the combination of this new cutting technique with methods for computer and robot assisted surgery is indispensable. A robotic assistance system which executes preoperatively planned cutting geometries using laser is not existent nowadays and is thereby proposed in the scope of this doctoral thesis.

1.1 Scientific Challenges

Fast and precise deflection of the laser beam using a beam deflector and positioning this device using a robot system are well known from industrial applications for cutting and welding material. The main challenge is to meet the laser incision even after numerous repositionings of the robot and to continue the cut exactly. However, methods used in industrial applications for robot assisted processing of material using laser are not straight forward adaptable to medical applications. The patient as an inexact described *workpiece* with his individual anatomy and pathology necessitates *one of a kind work.*

Planning methods for the ablation of hard tissue in situ using laser are not known nowadays. Even though the state of the art reveals physical, chemical and biological dependencies and interrelationships, cutting bony tissue using laser was regarded on the one hand theoretically and on the other hand phenomenologically until now. One has to notice that nowadays no comprehensive understanding of laser bone ablation and all influencing parameters is existent.

While industrial applications aim at maximizing performance and process speed, exact implementation together with physiological constraints are the main focus in surgical interventions. Furthermore, the integration of a robotic system into the operating theater requires to assess methods of risk management.

Overall, this leads to the scientific challenges and open questions which are dealt with in the scope of this doctoral thesis:

- How to realize laser cutting of bone with robotic methods?

- Which cutting accuracies are required and obtainable with a system for robot assisted laser osteotomy?

- How to plan hard tissue ablation preoperatively?

- Which parameters are essential for controlling and optimizing the ablation process?

- Which surgical disciplines profit from robot assisted laser osteotomy?

1.2 Structure

This thesis is organized as follows: In Chapter 2, the fundamentals of laser and its interaction mechanisms with matter are introduced. An overview of industrial laser applications is given with regard to computer and robotic assistance. Afterwards the mechanisms which drive pulsed laser ablation for tissue removal are described and picked up in the presentation of medical laser applications. In Chapter 3, the main concepts of computer and robot assisted surgery are introduced. In this context the state of the art of medical laser applications utilizing methods of computer and robotic assistance is reviewed. Based thereon, open questions and scientific challenges in the scope of this doctoral thesis are stated.

The following chapters (4-7) each deal with a specific question regarding robot assisted laser osteotomy and the methods developed within the scope of this doctoral thesis. In Chapter 4, the laser ablation process is considered as a discrete process and analyzed in order to understand influencing parameters. The development of an incision by concatenation of single laser pulses is explained. The next Chapter 5 deals with planning of robot assisted laser ablation. The methods developed for transferring a geometrically predefined cut into a sequence of laser pulses under consideration of the beam deflector's working area are introduced. In the following Chapter 6 the system realized in the scope of this doctoral thesis for robot assisted laser osteotomy is presented. Beside hardware and software components, the registration and calibration methods developed are introduced. Furthermore, the previous concepts and methods are put in the context of the overall workflow for robot assisted laser osteotomy. The corresponding risk analysis performed for the overall system is given in Annex A. In the last core Chapter 7 the experimental feasibility studies assessed in the scope of this doctoral thesis are presented.

The results of this doctoral thesis and all the innovations behind world's first robotic system for laser osteotomy are summarized in the last Chapter 8. The new concepts and methods are discussed with regard to their medical application. The doctoral thesis closes with an outlook on further research issues.

2

Laser Material Processing

Laser radiation was first created by Theodore H. Maiman 1960 at the Hughes Research Laboratory using synthetic ruby [Mai60]. The interaction of laser radiation with matter and its application in material processing is in the scope of research since that time.

In parallel the industrial and the medical research discovered application fields for laser. Only a few of them have significance in material processing [Bäu04]. While the industrial use of laser radiation is often accompanied by means of controlled movement or precise positioning using articulated arms and robots respectively, laser radiation in medical applications is mostly conducted manually by hand.

The use of laser radiation for material processing is introduced in this chapter, beginning with the laser fundamentals and laser beam characteristics. After an introduction of light interaction mechanisms, the general interaction effects of laser with matter are described. The usage of laser in industrial applications especially with regard to computer and robot assistance is reflected, since its previous and actual impact on computer and robot aided surgery. In the last part of this chapter the interaction of laser with biological tissue is explained in detail. Furthermore medical laser applications with focus on hard tissue ablation are introduced.

2.1 Laser Fundamentals

The acronym *laser* for light amplification by stimulated emission of radiation is used for the principle of operation and for devices, emitting electromagnetic radiation (cp. Figure 2.1) of one wavelength or several narrow bandwidth wavelengths. The light emitted is coherent and unidirectional. It can easily be collimated in order to achieve a focused monochromatic beam with small divergence.

A laser beam originates from an energy change between an excited and a lower energy stage. An incoming photon is absorbed by an atom or molecule, resulting in an excited energy state of the atom or molecule. Stimulated emission is then caused by a photon meeting an excited atom or molecule and thereby stimulating it to emit the stored energy in term of a photon, returning in an unexcited (lower) stage. The emitted photon is found to be in the same phase as the stimulating photon, also traveling in the same direction. Stimulated emission of photons can be considered as the opposite process of absorption.

Numerous materials can be utilized for stimulated emission (solid-state, gas, fluid, semiconductor). In order to emit light a population inversion of atoms or molecules in an excited energetic state in the active medium is necessary. This is realized by pumping the atoms or molecules optically or electrically. Photons passing through accumulated excited atoms or molecules stimulate the generation of numerous

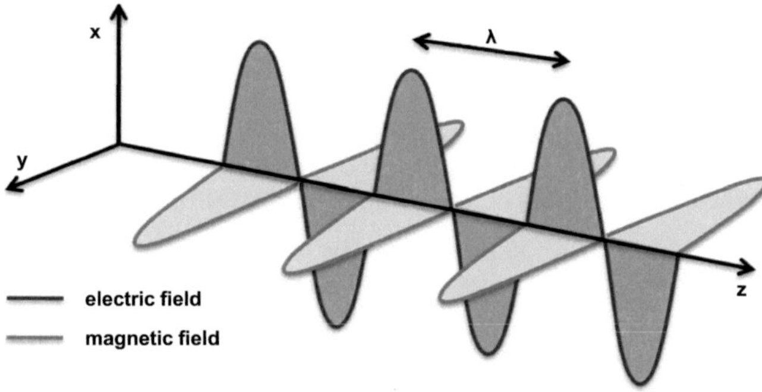

Figure 2.1: Scheme of an electromagnetic wave. The wave is characterized by the wavelength λ and the amplitude. The electric wave and the magnetic wave are perpendicular and propagating in z direction.

photons, each of them also stimulating excited atoms to emit photons. The amplification is caused by stimulated emission. Figure 2.2 illustrates the basic principle of a laser system.

Laser light can be classified by its Transverse Electromagnetic Mode pattern (TEM), which gives the number of radial, angular and longitudinal zero fields: $TEM_{\rho\varphi q}$. The index q of the longitudinal zero fields is normally not used. Most lasers for material processing are used in TEM_{00} or TEM_{01}. The higher the laser mode, the harder it is to focus the laser beam to a fine spot.

Furthermore laser light is characterized by polarization, i.e. the electric vectors of all waves are lined up. General laser light without any filtering is randomly polarized. Dependent on the polarization filter used a distinction is drawn between linear, circular and elliptical polarization [Sil04].

A laser beam is highly directive and therefore shows only very small divergence. Propagating in free space and using a lens a Gaussian beam (TEM_{00}) shows the minimum spot size $w_{\sigma 0}$ on its beam axis z at

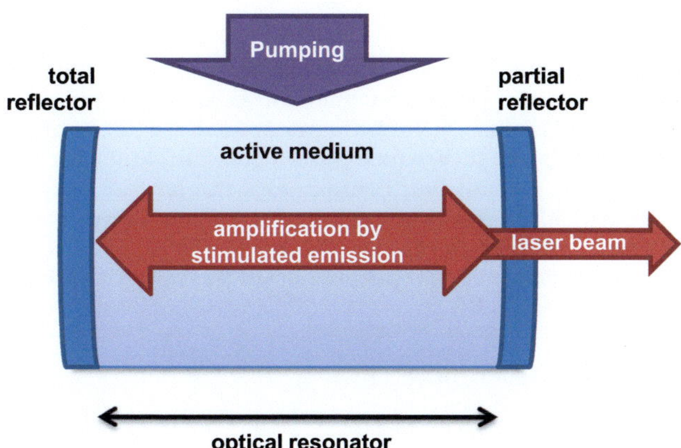

Figure 2.2: An optical resonator containing an active medium providing suitable energy stages is stimulated by a pumping source in order to create a population inversion

the beam waist (cp. Figure 2.3). The variation of the spot size for this beam is given by

$$w(z) = w_{\sigma 0}\sqrt{1 + \left(\frac{z}{z_R}\right)^2}.$$

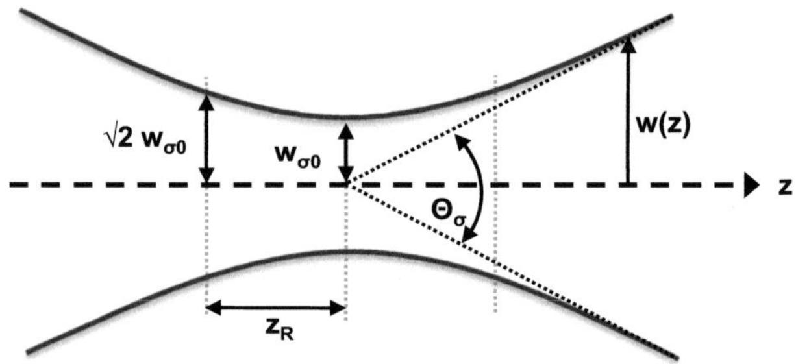

Figure 2.3: Notation for Gaussian beam diverging from its waist.

Without loss of generality the origin of the z-axis coincides with the beam waist, where the Rayleigh range z_R is defined depending on the wavelength λ

$$z_R = \frac{\pi w_{\sigma 0}^2}{\lambda}.$$

The Rayleigh range is the distance along the path of propagation from the beam waist to the plane in which the beam diameter exceeds the waist diameter by factor $\sqrt{2}$. It characterizes the near field of the beam, while the far field beam divergence gives a constant angle for the expansion of the beam. A Gaussian beam shows the largest Rayleigh range and lowest far field divergence compared to higher Transverse Electromagnetic Modes [Ion05].

The depth of focus (confocal parameter) of the beam is defined as

$$b = 2z_R,$$

where spot size of the beam reaches

$$w\left(\pm z_R\right) = \sqrt{2}w_{\sigma 0}.$$

The beam quality factor M^2 indicates the quality of the laser beam. A high beam quality means that the laser beam has a small focus with high intensity and small divergence. Nowadays, the term beam propagation factor and the notation K are used in almost the same manner.

For a theoretically ideal Gaussian beam $M^2 = 1$, while for real beams $M^2 > 1$. Hence, the beam quality factor is defined as:

$$M^2 = \frac{1}{K} = \frac{\pi}{\lambda} \cdot \frac{w_{\sigma 0} \Theta_\sigma}{2}$$

with the divergence angle Θ_σ. M^2 describes the ratio of divergence between the considered beam and that of theoretical diffraction-limited Gaussian beam with the same waist.

Among the theoretical determination of beam parameters, the experimental validation is often crucial in order to accurately describe the beam divergence for a specific laser setup. Most commonly knife-edge scanning techniques and derivate methods are used to measure the radius of a Gaussian beam [Arn71, Fir77, Joh98, Kla83]. Ronchi ruling is another method for determining the Gaussian beam diameter, which allows measuring very small beam diameters [Che03].

Furthermore the intensity distribution of the laser beam is of high interest in material processing. Numerous methods are proposed, to analyze the beam profile of a laser beam and mostly image processing techniques are utilized these days [Als06, Sch81].

2.2 Light Matter Interaction Mechanisms

When light interacts with matter, the matter responds in a proportionate way. Reflection, refraction, scattering and absorption are counted among these effects, which constitute the basis for laser material processing.

2.2.1 Reflection and Refraction

Reflection and Refraction are strongly related to each other (see Figure 2.4). Refraction is the change in the direction of the light wave when passing from one media into another. Passing from material M_1 with refraction index n_1 to material M_2 with n_2 the beam's inclincation angle with respect to the surface normal changes from φ to φ''. This change in direction is induced by the difference in the speed of light traveling through the two media. The refraction index of a certain material defines the ratio of the speed of light in the material to the speed of light in a vacuum.

Reflection occurs whenever incident light strikes a physical boundary between two materials of different refraction indices (e.g. air or water). The reflection angle φ' equals the angle of incidence φ. The angles are measured between the surface normal and the incident respectably reflected beam.

If the surface itself is smooth, with small irregularities in comparison to the wavelength of the incident beam, the reflection is called *specular* and the reflected beam lies within the plane of incidence. While against that a rough surface, comparable or even larger than the wavelength of radiation causes *diffuse* reflection. In this case several beams are reflected which do not necessarily lie within the plane of incidence.

2.2.2 Scattering and Absorption

Light scattering and absorption are the two important processes contributing to the visual appearance of objects. Scattering changes the propagation direction of the electromagnetic wave due to differing refraction indices inside the medium (cells, particles, etc.).

Absorption of electromagnetic radiation characterizes the uptake of energy by a medium. The inverse of the absorption coefficient μ_α defines the distance after which the intensity has decreased by $1/e$ from its incident value. Corresponding to this, the relation between the light intensity and the depth of indentation is exponentially decreasing.

Absorption is strongly dependent on the target material and the wavelength of the electromagnetic radiation. Figure 2.5 shows the absorption coefficient of water in dependency of the wavelength.

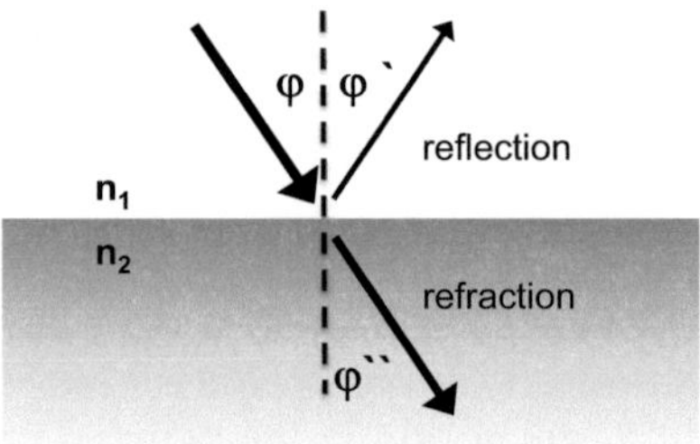

Figure 2.4: Geometry of specular reflection and refraction.

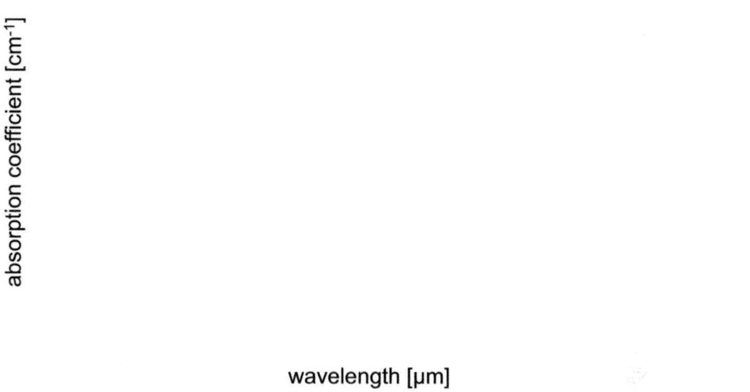

Figure 2.5: Absorption coefficient μ_α of water plotted against the wavelength λ. Graph generated with data from [Ore09].

2.3 Absorption Processes in Matter Caused by Laser Radiation

As previously described, when electromagnetic radiation strikes a surface interaction effects occur. Some of the radiation is reflected, some is transmitted and some is absorbed. The absorption of the radiation is the most important phenomenon in laser material processing, whether in industrial or medical applications. On its way through the target medium the radiation is absorbed following different rules of absorption, which are dependent on the wavelength and intensity as well as on the medium itself (cp. Section 2.2.2).

Absorption results in various effects such as heating, melting, boiling etc., which are the basis for several laser material processing techniques [Ste03a]. Figure 2.6 visualizes several of these effects.

By irradiating the surface of a material with a laser beam, electrons of the material are excited. This excitation energy is converted into heat. Depending on the thermal properties of the material and the laser parameters a temperature distribution is established. If the incident laser intensity is sufficiently high, the laser energy absorption causes phase transformations. The application of external energy leads to in-

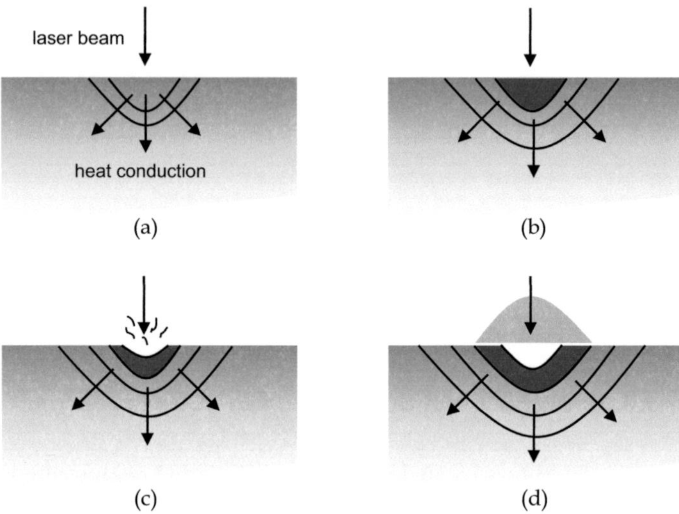

Figure 2.6: Laser material interaction effects: (a) heating, (b) surface melting, (c) surface vaporization, and (d) plasma formation.

creasing molecular vibration. Surpassing the material specific melting threshold the phase change to liquid occurs, since the materials molecules become less ordered. The depth of melting cannot be infinitely high, since once the increasing temperature at the surface of the irradiated material reaches the boiling point, the maximum melting depth is reached. With reaching the boiling point, the vaporization of the material takes place, i.e. the temperature at which the vapor pressure of the liquid is equal to the external pressure. Once the vaporization began, the liquid-vapor border proceeds inside the material under continuous laser irradiation and material is removed by evaporation simultaneously [Dah08].

Significant surface evaporation occurs, if the material is irradiated with a sufficiently large laser intensity. Hence interaction between the incident laser beam and the vapor takes place, which influences or interferes the overall laser interaction process. Most important thereby is the ionization of the vapor, i.e. the formation of plasma.

2.4 Industrial Laser Applications

Since the laser invention many hundred laser systems have been developed, but only a few of them are of commercial significance in industrial applications. Since the irradiation of a material with the electromagnetic energy of a laser beam induces thermal or thermo-mechanical response, industrial applications utilizing laser always rely on the conversion of energy. The focused laser beam is one of the highest power density sources available for industrial applications these days [Ste03a]. Furthermore the possibility to shape this power in time and space is almost unique [Ste03b].

Generally one can say that the increasing application of laser in material processing is due to several advantages which are unique for this energy source, e.g. non-contact processing, improved quality, higher productivity, automation worthiness. Laser material processing is a wide field and several applications are existing. Following Majumdar et al. one can distinguish applications which require limited energy causing no significant change of phase or state in the material and applications which require a substantial amount of energy inducing phase transformations [Maj03]. The last-mentioned class of applications can be further subdivided into joining and machining applications [Dub07], which are regarded in the scope of this thesis, especially cutting, drilling and welding. Figure 2.7 illustrates laser material processing applications in dependence of the interaction time and laser power density. The laser power density and interaction length are carefully suited for each process regarding the material to be processed and its certain parameters. Generally, processing material necessitates the deep understanding of the parameters influencing the process. Therefore numerous models for laser processes were developed and evaluated [Yil97a, Yil98, StO02, Yao05].

After laser cutting found its way into industry, laser welding became a promising application field. The high processing speeds in laser applications and small tolerances require a high degree of automation. Since the invention of the laser there has also been a constant development towards shorter pulse durations [Mei04]. Nowadays pulses in femtosecond range are obtainable and even shorter, allowing to input high laser power densities in very short time and therefore opening up new possibilities in laser material processing. In the following laser cutting, welding, drilling and micro-structuring are further explained.

2.4.1 Laser Cutting

Today laser cutting is the most common industrial application of laser. In comparison with alternative techniques the laser cuts faster and with a higher quality. Hence, the process of laser cutting is characterized by contact freeness and easy automatability. The advantages of laser cutting against competing processes (e.g. water jet cutting, conventional stamping) are [Ste03a]:

- Very narrow cutting kerf (width of cut opening)

- Cut edge can be clean and smooth

- No edge burr

- Very narrow heat affected zone

The induced laser power heats the material to the melting temperature. A gas jet accelerates the molten material, which is thereby separated from the material and the cutting kerf evolves. Laser cutting found its way into the automotive industry which requires the processing of structural or chassis parts, where complex end-parts are build

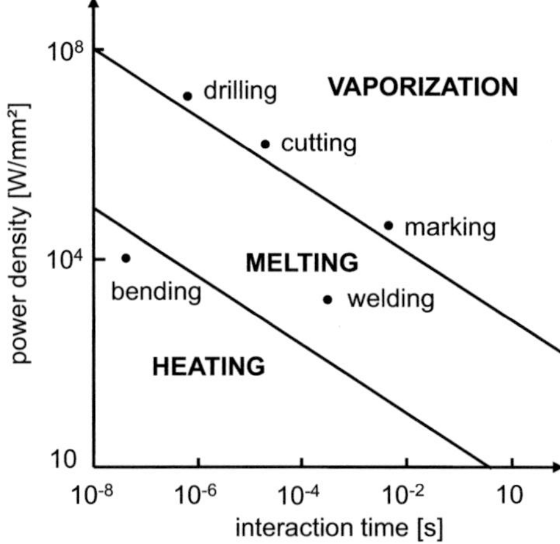

Figure 2.7: Schematic view of laser material processing applications in dependence on the laser power density and the interaction time (Adapted from [Maj03]).

14

using metallic sheets with thicknesses below 3 mm. Driven by the demand for cutting even thicker metal plates, increasing laser powers are applied. Laser cutting parameter optimization plays an important role and is in the scope of current research [Dix06, Jim07, Ari07, Keu07, Tay94a]. However, the possible parameter variance is more narrow for achieving good quality laser cuts of thick metal plates, since a critical balance between the process parameters has to be considered (e.g. laser power, focal distance, cutting velocity and gas pressure). Therefore cutting of thick plates is nowadays not that highly automated as laser cutting of thin plates and necessitates human supervision.

Maintaining the correct beam focus at the workpiece is the key requirement for laser cutting. While this is simple for flat metal sheets, it is far more challenging considering 3d freeform workpieces. Such applications require active control of the focal position [Bog08]. This can either be achieved using a positioning table for the workpiece and apply Computer Numerical Control (CNC) or by using robots to position the laser cutting head [Roo00].

Dworkowsky et al. [Dwo95] proposed an x- and y- gantry positioner in order to move the cutting head along a pre-programmed path for processing the laser cut. Optical beam delivery is realized comparable to a two-dimensional scan head, since the light-weight mirrors are mounted to the x- and y-axis of the gantry. A lens is utilized to focus the scanned laser beam onto the workpiece. The system was designed to achieve high-speed precision movements and patented. However, Dworkowsky et al. did not explain the positioning and alignment of the workpiece in respect to the gantry. Another x-y-motion table for positioning the workpiece in respect to the laser head was proposed by Hace et al. [Hac97].

Several special robot systems were proposed for laser cutting applications [Wan99, Cho00]. Bruzzone et al. proposed a mechanical structure for a parallel robot for laser cutting applications [Bru02]. They motivated their approach by the excellent dynamic performance, high stiffness and accuracy reachable with parallel robots. However, these systems never left the state of mechanical design. Usually conventional six degree of freedom serial robot arms are utilized in industrial applications.

Jenoptik Automation presented the next generation of robot based CO_2 laser systems Votan C-BIM (Beam in Motion) for laser cutting and trimming of metal sheets. The system allows up to 5 kW of laser power with high accuracy and dynamics [Jen09]. The laser is mechanically decoupled from the robot and the complete optical path of the beam is integrated into the robot itself by coupling it into the foot. Each robot axis contains a water-cooled mirror at the turning point, thereby delivering the laser beam to the cutting head. The reflection loss over the complete optical path is stated to be lower than 5 %. A Stäubli TX90 robot was specifically adapted for this application (see Figure 2.8).

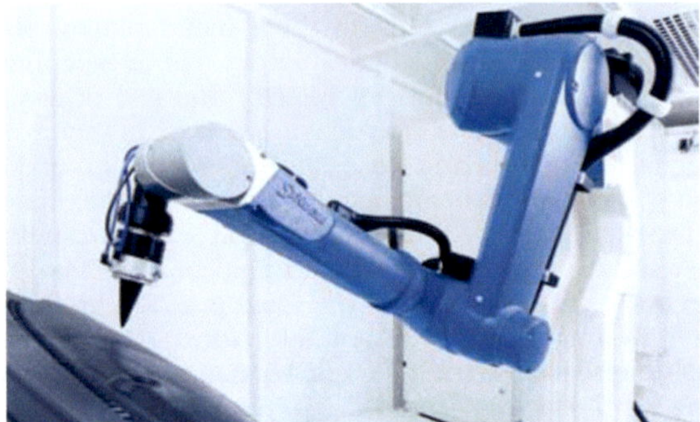

Figure 2.8: Next generation of robot-based CO_2 laser systems Votan C-BIM (Beam in Motion) with incorporated optical path. (Picture: JENOPTIK Automatisierungstechnik GmbH.)

Mostly programming of laser cutting robots (and industrial robots in general) is performed manually using a teach-pendant or offline using a virtual scene. Using a dynamic programming procedure Pashkevich et al. developed multiobjective optimization methods for robot motion considering kinematic redundancy and particularities of laser cutting [Pas04].

Furthermore several studies were applied in order to improve laser cutting by means of online process control. Roggero et al. proposed a classification system for analyzing the quality of laser cuts on the fly and controlling of the process [Rog01]. A camera is applied, which surveys the process. The image sequence is synchronized with the state information of the cut and then classified using a neural network.

2.4.2 Laser Welding

After laser cutting also laser welding became a state of the art application in the industrial area. The aim of the laser welding process is to create a pool of molten material (weld pool) at the overlapping work-piece surfaces [Dah08]. Almost every material can be laser welded. The challenges are in beam and material handling.

For laser beam welding a laser beam is focused on the workpiece, energy is deposited at the surface and then transported deeply inside a narrow cavity by heat conduction. Again the feature to concentrate a very high energy in a very small area characterizes the laser induced process. If the power density exceeds a certain threshold (> 10^6 W mm^{-2}), key holing in the weld pool occurs [Ama02]. The evaporation induces recoil pressure form this small depression which develops into the keyhole by the upward displacement of molten material along the keyhole walls [Dah08]. Within the keyhole the laser energy is multiply reflected and therefore is of high importance in order to achieve deep penetration of the laser beam into the material. The small heat-affected zone by the laser thanks to high power densities and the fast cooling rate the process induces only low heat to the material.

During laser welding the surface of the material undergoes rapid vaporization and the subsequent ionization of the vapor creates a standing plasma above the surface of the material. Thereby laser radiation is absorbed and scattered, what impedes effective welding. Therefore various shielding gases such as argon, helium or CO_2 are utilized. Simultaneously to the shielding gas also hardfacing material (e.g. in form of a powder) can be introduced into the laser beam and thereby melted.

Understanding the welding process is of high importance in order to optimize the process and efficiency [Yil97b, Tso06]. As for laser cutting also in the scope of laser welding various models were applied [Ama02]. Furthermore constant development in the scope of beam scanning towards more accurate, faster and smaller systems can be observed. For example, Schmitt et al. recently proposed a miniaturized laser processing optics for micro welding [Sch08].

Robotic Laser Welding

Industrial robotic welding is a popular application of robotics worldwide, especially in car industry [Pir02]. The high welding speeds in laser welding, e.g. 100 mm/s and the small tolerances require a high degree of automation [Mei04]. Most research work considering robotic laser welding is conducted in the field of open loop control considering the influence of various control variables and quality parameters.

Programming for robotic laser welding is mostly performed offline [Wai07]. Combining off-line programming, motion planning and sensor-based control of the robotic welding process is combined by Madsen et al. [Mad02]. For closed-loop or adaptive control mostly CCD cameras are utilized, which are used to determine the position of the focal point for example. Online focus control sensors incorporated into the laser delivery system, i.e. an optical fiber in this case, measuring the process light returning through the delivery system was proposed by [Han99].

Andersen developed a feed forward control strategy based on gain scheduling [And01]. Measuring the actual gap between workpieces with a laser sensor, the control parameters are adapted accordingly. The welding speed, wire feed rate and position of the laser focus are considered in the control parameters. Seam tracking to control the robot movements using a laser scanner, like the work proposed by Fridenfalk et al. [Fri03], Xu et al. [Xu04] or de Graaf [Gra07], allows to continuously correct both position and orientation of the welding torch at the robots end-effector along the seam. Therefore necessity of trajectory programming or geometrical databases may be replaced by online adaptive control of the robotic welding system. Bolmsjö et al. pointed out that the use of sensors controlling the welding process itself and the robot action are of primary interest [Bol05].

Remote Laser Welding

Thanks to high power laser development with high beam quality, long focal lengths with small spot sizes are achievable and allow to utilize focused laser beams with a comparable long Rayleigh length. In combination with highly dynamic beam deflection long focal lengths are characteristic for remote welding systems. While in conventional laser processing the work head is close to the workpiece, a standoff distance allows to use the laser with its non-contact processing feasibility. Hence, the laser beam (with $f \geq 1000\,\text{mm}$) is focused onto the workpiece [Bäu04] and deflected by x-y-scanning systems. Consequently the beam deflection unit needs to be positioned in respect to the workpiece with high accuracy and repeatability. Key benefit of the standoff distance and the ability to maneuver the laser is the comparable larger working area of remote welding systems. Hence, remote laser welding systems usually comprise a high power laser source, a beam delivery and deflection system as well as a serial robot with six degrees of freedom. The robot positions the beam deflection unit in respect to the workpiece in order to process welding spots or seams [Bey03]. Mostly Nd:YAG and CO_2 laser are utilized for remote laser welding systems [Tso08]. Figure 2.9 illustrates remote welding in automotive industry.

The development of remote laser welding is mainly driven by the automotive industry in order to increase productivity and construction strength and to enable new designs. Compared to traditional laser welding with short focal length optics, high speed movement of the laser beam in remote laser welding significantly contributes to the reduction

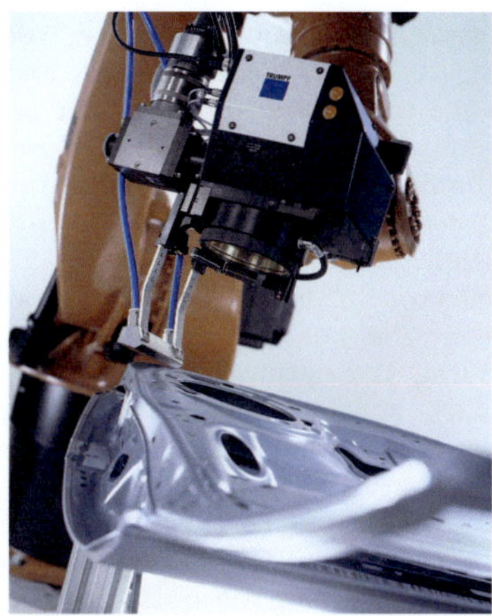

Figure 2.9: Remote laser welding with a robot positioning a three-dimensional scanning device relative to a workpiece. (Picture: TRUMPF GmbH + Co. KG.)

of cycle times and the accessibility to areas not accessible with short focal lengths. Hence, remote laser welding promises a wider application spectrum. However, the main challenge for remote welding systems is part handling, since the processing time is so fast [Koc02]. Typically 20-50 % of the welding process cycle time are only used for the welding, while the positioning and loading time of the systems is consuming the remaining time [Gru03].

A variety of publications deal with planning and optimization of remote laser welding, since this technique promises higher productivity in comparison with conventional welding techniques. Stemmann et al. modeled the scanning unit mounted to the robots flange as a kinematic chain accounting for the redundancy resulting from the combination of two actuated devices [Ste06]. The task of sequential welding is related to the Generalized Travelling Salesman Problem (GTSP) which is solved in order to find the time optimal path. The solution was applied to a simulation environment and the authors state that 20% of the execution time could be saved. However, the method was not applied to a real application.

Reinhart et al. proposed a task based programming system for remote laser welding which is capable of automatically transferring task descriptions into executable robot operations [Rei08]. The system does not utilize a scanning device, in fact a conventional long range optic is directly attached as the robots end-effector. The user moves a tracked 3d-input stylus across the surface of the workpiece and a projector is then used to augment a virtual cursor onto the workpiece directly. Thereby the user can define a trajectory on the surface by simply drawing it (cp. Figure 2.10a).

The generated task can be divided into seam and point level. For optimizing the path for the welding task, the problem is divided into two subproblems. First, the optimal welding sequence is determined automatically under the assumption that the shortest Cartesian path between all seams results in the shortest cycle time for the welding problem. This problem is an instance of the Traveling Salesman Problem (TSP) and solved by a simple heuristic approach (cp. Figure 2.10b). Second, the variation of inclination angles is regarded under the assumption that a smooth path of the optic results in improved robot movements regarding the cycle time by avoiding main axis oscillation.

In order to increase the positioning accuracy of the scanner unit by the robot, several online sensing methods are applied. Huisson et al. introduced a calibration methodology for a seam tracking sensor to position the laser focal point and to modify the tool trajectory [Hui02]. They combined a laser profiling sensor and a laser focal point sensor based on a CCD camera, but the overall accuracy achieved in the calibration procedure did not reach the range of accuracy Huisson et al. claimed for robotic laser welding. Seam tracking based on stereo visual feedback control was proposed by Zhang et al. [Zha07] in order to improve the dynamic trajectory accuracy. Visual measuring of the path accuracy for robotic laser welding was also in the scope of the work performed by Du et al. [Du04].

2.4.3 Laser Drilling

Drilling applications represent only 5% of the industrial laser material processing applications. In laser beam drilling four types are distinguished: single pulse or percussion drilling, trepanning or helical drilling. In single pulse drilling holes with a diameter of 0.015 mm to 1.2 mm can be achieved using lamp-pumped pulsed industrial Nd:YAG solid-state lasers. In trepan drilling, the laser beam is guided along the circumference of the hole which will be generated, while percussion drilling directly removes the material. Helical drilling is applied when cylindrical or conical shaped holes need to be achieved and geometrical

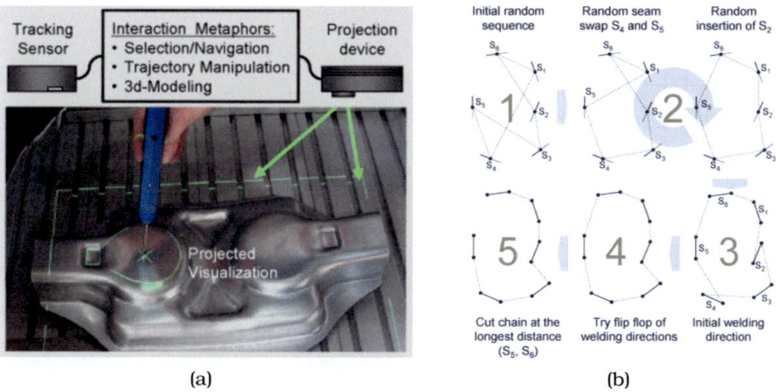

Figure 2.10: Task based programming system for remote laser welding. (a) User defines a trajectory on the workpiece surface by simply drawing it supported by augmented reality.
(b) Finding the optimal welding sequence by referring it to
the Traveling Salesman Problem. (Reprinted from [Rei08],
Copyright 2008, with permission from Elsevier.)

tolerances below 2% are required. Further reading and exemplary applications can be found in the literature [Dah08, Ste03, Che07, Boo04].

2.4.4 Laser Micro Structuring

Defined material removal for micro structuring is another industrial
application field of the laser. Here the laser is often used in pulsed
mode with high intensities in order to quickly heat up the material
and induce removal. To gain high intensities mostly solid state lasers
are utilized in TEM_{00} mode and beam quality factor near $M^2 = 1$ to
achieve good focusing abilities. For deflecting the laser beam, in this
application field galvanometric scanner systems are utilized too. In this
scope the shorter the laser pulse is with the same intensity, the better
the quality of the resulting crater of removed material by one single
laser pulse is [Kor07]. Figure 2.11 illustrates the laser ablation process
by sublimation. A short laser pulse applied to a workpiece heats it
at the incidence point to the melting point. Instantly a plasma plume
occurs due to ionization of ablation products. Increasing pressure by

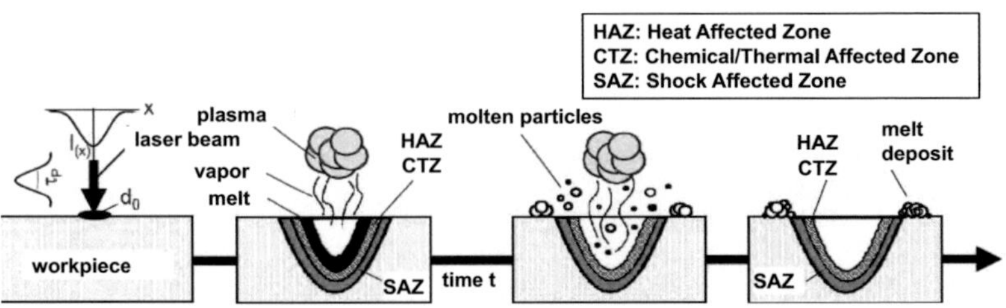

Figure 2.11: Phases of laser induced material removal. (Adapted from [Kor07].)

vapor leads to evaporation of molten particles. Molten particles solidify again on the surface around the crater margin.

For micro structuring pulse length in femtosecond and picosecond range are usual [Gil05]. Mostly single atoms are released out of their matrix using femtosecond pulses. Since energy deposition is faster then heat dispersion, nearly complete sublimation occurs, which is also called cold ablation. In contrast pulses in picosecond range already show heat dispersion in the adjacent material and first surface border influences occur. Furthermore femtosecond and picosecond pulses only remove small amounts of material, allowing high accuracies. The logarithmic dependency of ablation depth from the absorbed laser energy can be determined accurately for industrial materials (e.g. copper) [Kor07].

2.4.5 Conclusion

The utilization of laser for industrial applications is well established and a method of choice nowadays. One the one hand, this is due to the fact that laser allows for a high degree of automation and on the other hand, that the processing quality can be enhanced. However, with respect to the medical application area, the well established methods cannot be applied offhand. In fact, treating biological tissue with laser necessitates a deeper look to the interaction principles under consideration of the tissue properties. The next section reviews laser interaction processes and the impact on biological tissue. Afterwards medical applications using laser are introduced.

2.5 Interaction of Laser Radiation with Biological Tissue

Interaction of laser radiation with biological tissue is mainly based on absorption. The energy induced by the laser beam into the tissue is converted into heat. The photothermal event results in warming, coagulation or excision, or incision through tissue vaporization. Scattering of the laser beam is only of importance regarding deep tissue penetration [Cob06].

Absorption in biological tissue is dependent on the following parameters:

- wavelength

- power or energy

- continuous or pulsed waveform (cw/pw)

- exposure time or pulse duration

- power density (spot size)

- optical properties

Depending on the wavelength important absorbers in the ultraviolet range of the spectrum are peptides and nucleid acids. In the visual range mostly carotene, melanin and hemoglobin are absorbers. Longer wavelengths in the infrared range are mainly absorbed by water and hydroxyapatite, both crucial factors for laser bone ablation. The optical properties of the tissue are characterized by water content, pigmentation, mineral content, heat capacity (i.e. thermal conductivity and tissue density) and latent heat of transformation (i.e. vaporization water, denaturation of proteins, melting minerals).

The interaction effect strongly depends on the power density and the exposure time. The number of combinations of exposure parameters is almost unlimited. Nevertheless the interaction of laser with biological tissue can be categorized into five areas with respect to the exposure time and power density: photochemical, thermal, photoablation, plasma induced and photodisruption. The following subsections and Figure 2.12 explicate these laser-tissue interactions.

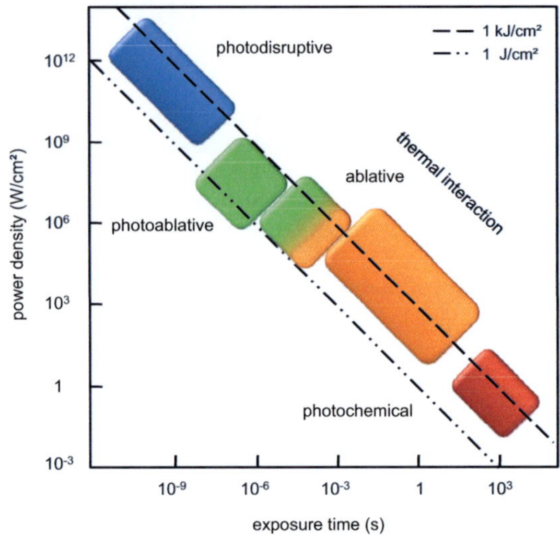

Figure 2.12: Scheme of laser-tissue interactions based on Boulnois [Bou86]. Exposure times and power densities used in biological applications vary over wide ranges. The slashed and dotted lines indicate the energy density.

2.5.1 Photochemical Interaction

Using very low power densities ($1\,\mathrm{W/cm^2}$) and long exposure time (continuous wave or exposure times longer than one second) induce chemical reactions of the laser within the irradiated tissue or macromolecules. Thanks to their high optical penetration depths and efficiency lasers with wavelengths in the visible spectrum are mainly utilized to induce photochemical processes.

One important medical application, which utilizes photochemical interaction is the photodynamic therapy (PDT), where a photosensitizer is injected, distributes among the tissue and remains inactive until irradiation. In healthy cells the clearance is faster than in the latter cells. This characteristic is utilized in order to irradiate tumor cells in which the photosensitizer is still concentrated days after injection.

2.5.2 Thermal Interaction

Thermal interactions build the largest group of tissue interaction mechanisms. Through the conversion of electromagnetic energy into thermal energy, the temperature in biological tissue increases under laser irradiation and cause thermal effects. They are dependent on duration and peak value of the tissue temperature achieved. Exposure times from 1 min down to 1 µs with power densities from 10 to 10^6 W/cm^2 are typical parameters indicating thermal interactions.

The effects of heat on biological tissue are characterized by the temperature and summarized in Table 2.1. Until a temperature of approximately 37 °C no damage is caused. A temperature increasing about 45 °C leads to changes in the cellular membranes, which can induce hyperthermia. If hyperthermia lasts for several minutes, a significant percentage of the tissue becomes necrotic [Nie07]. Exceeding 50 °C the enzyme activity is reduced, which in turn results in a reduced energy transfer and the immobility of cells. Furthermore self-repairing mechanisms are inhibited and the percentage of cells which survive the temperature increase is further reduced. Beyond 60 °C proteins and collagen of the extra-cellular matrix denature. Hence the tissue becomes coagulated and necrosis of the cells takes place. Coagulated tissue is characterized by its darker appearance in comparison to healthy surrounding tissue. The coagulation effect is often utilized in surgical cutting of soft-tissue for hemostasis (laser scalpel). Furthermore malignant tissue (e.g. tumor) can be treated with coagulation in order to kill the

Table 2.1: Thermal interactions of laser radiation with biological tissue characterized by temperature. [Nie07]

Temperature (°C)	Thermal effect
37 °C	no irreversible damage
45 °C	hyperthermia
50 °C	reduction of enzyme activity, cell immobility
60 °C	denaturation of proteins and collagen, coagulation and necrosis
80 °C	membranes become permeable
100 °C	vaporization, thermal decomposition (ablation)
>100 °C	carbonization
>300 °C	melting

affected cells and stimulate the replacement with healthy tissue during the healing process.

Increasing the temperature above 80 °C, causes the membranes to become permeable so that the equilibrium of chemical concentrations cannot be maintained. At 100 °C the vaporization of water contained in the irradiated tissue starts. The temperature of the exposed layer remains at approximately 100 °C while the continuing irradiation dehydrates the tissue [Ber03]. The removal of water decreases the local thermal conductivity and thereby the heat conduction to the surrounding tissue is reduced. This phase transition from liquid to vapor inherits a large increase of volume. The generated gas bubbles induce mechanical rupture and thermal decomposition of tissue fragments. This thermal decomposition is also called thermo-mechanical ablation due to the significant build-up of internal pressure [Nie07]. At this time the temperature is further increasing and exceeds 100 °C, where carbonization of the tissue starts. It can easily be visually observed since a characteristic blackening of the tissue occurs and smoke develops. Finally with a temperature beyond 300 °C the material may begin to melt.

If radiation enters homogeneous tissues, the intensity decreases with increasing penetration depth. A proportional temperature gradient develops inside the tissue. In the region with more than 100 °C the tissue vaporizes but a thin layer of carbonization remains. The following zone is coagulated. These changes are irreversible. Furthermore a region, where the tissue was slightly warmed up, remains in which hyperthermia occurs, i.e. reversible. Figure 2.13 schematically indicates the thermal effects for biological tissue corresponding to the temperature gradient. The extend of a certain zone depends on the penetration

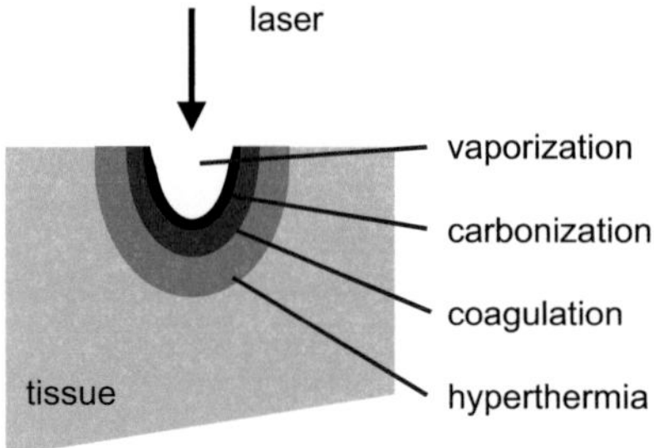

Figure 2.13: Thermal effects and their location for laser irradiated biological tissue.

depth of the laser wavelength, power, penetration time (cw or pulsed) on the one side, and on the other side on the tissue properties.

2.5.3 Photoablation

The direct breaking of molecular bounds by high energy photons is called photoablation. This process is non-thermal, since it proceeds much faster than vaporization [Ber03]. Nevertheless the process is of a thermal nature, but the thermal influence on the remaining tissue is negligible.

2.5.4 Plasma-Induced Ablation

Power densities exceeding 10^{11} W/cm^2 lead the to optical breakdown in solids and fluids. The ionized plasma induces a very clean ablation, associated with audible report and blueish plasma sparkling. Pulse durations therefore are typically varying from 100 fs to 500 ps.

2.5.5 Photodisruption

High power densities from 10^{11} W/cm^2 to 10^{16} W/cm^2 and extremely short pulse durations from 100 fs to 100 ns lead to fragmentation and cutting of tissue by mechanical forces. This interaction is also called photomechanical, because it is more or less based on the direct transfer of optical into mechanical energy. Whereas thermal processes can also be involved, the overall process is considered to be athermal regarding the remaining tissue.

2.6 Pulsed Laser Ablation

In the scope of this doctoral thesis the process of pulsed laser ablation (with pulse durations lower than 1 ms) is especially of importance in order to process bony and cartilage tissue. The mechanisms driving ablation of biological tissue are depending on several properties which can vary from tissue to tissue and individual to individual. Therefore it is important to understand the processes induced by irradiation with pulsed laser in a comprehensive way. While the objective of this thesis is the combination of pulsed laser processing with robotic assistance in the medical area, which is by definition an interdisciplinary research area, describing the mechanisms of pulsed laser ablation itself necessitates physical, chemical and biological knowledge combined with mathematical understanding. In this section a comprehensive overview of pulsed laser ablation is given, which is mainly based on the summary review of Vogel et al. [Vog03]. Further citations are quoted directly.

2.6.1 Biological Tissue Properties

Pulsed laser ablation is a complex process which is dependent on the properties of the irradiated biological tissue. First of all the optical properties which are established by the composition and morphology of the tissue are influencing the volumetric energy distribution which is mainly responsible for the ablation process. Furthermore the thermo-mechanical response of tissue to pulsed laser radiation is characterized by its mechanical and structural properties.

Optical Properties

The cells of biological tissue are embedded into a great amount of extracellular material: the extracellular matrix (ECM). The ECM is a complex and composite material, which maintains the structural integrity of the tissue. Molecular components of the ECM are proteoglycans, Non-proteoglycan polysaccharide, fibers (collagen and elastin), etc. The ratio of components of the ECM varies significantly among the tissue types. Distinction is made between cell-continuous and matrix-continuous tissue. Cell-continuous tissue only has a little ECM and collagen content (e.g. liver), while matrix-continuous tissue is characterized by small cellular fraction and high amount of collagen (e.g. dermis, cartilage). The ECM of bone contains mainly collagen (about 90%) and mineral deposits (e.g. calcium phosphate, magnesium, carbonate). The interaction of collagen and further ECM components with water significantly influences the energy transport mechanisms of the tissue. For most wavelengths only a single tissue constituent, like water or collagen, absorbs the radiation and characterizes the interaction process.

Mechanical Properties

Ablation process kinetics and dynamics are modulated by the elasticity and strength of the tissue. The strength of a specific tissue is positively correlated to its collagen content. Together with extensibility the stresses and deformations necessary in order to achieve material removal by fracturing the tissue matrix, as for laser ablation can be derived.

Thermal Denaturation

Thermal denaturation of tissue is the process which transforms the tissue from a native to a denatured state. This process does not solely depend on the temperature but also on the exposure duration [Wel95]. The thermal injury of a tissue can be estimated by applying the Arrhenius rate integral. For a given thermal transient $T(t)$ the accumulation of thermal injury can be expressed as

$$\Gamma(t) = \int_0^t A \exp\left(\frac{-\Delta E}{k_B T(t')}\right) dt',$$

with the Boltzmann constant k_B, the activation energy ΔE and the frequency factor A. $\Gamma(t)$ is a dimensionless measure of the degree of thermal injury accumulated at time t [Wel95].

For pulsed laser ablation the temperature required to affect the mechanical stability of the irradiated tissue exceeds $100\,°C$.

2.6.2 Energy Response

The laser ablation process is mainly driven by the spatial distribution of the volumetric energy density. This is controlled by the incident radiant exposure Φ_0, the optical absorption and scattering properties of the tissue.

Optical Absorption Properties

The optical absorption properties of tissue are dominated by the absorption of proteins, DNA, melanin, hemoglobin and water (see Figure 2.14). The optical absorption T is expressed by Beer-Lambert law:

$$T = \frac{I}{I_0\,(1 - R_S)} = 10^{\epsilon c l} = \exp\left(-\mu_\alpha l\right),$$

with the specular reflection R_S, the radiant exposure transmitted I_0 and after traveling through I an optical path of length l in a sample with molar extinction coefficient $\epsilon\,(mol^{-1}cm^{-1})$ and concentration $c(mol)$. Generally the absorption properties of a tissue are expressed by the absorption coefficient $\mu_\alpha\,(cm^{-1})$.

In the infrared (780 nm to 15 µm) water is the most important tissue component that contributes to absorption with $\lambda \geq 900$ nm. While the impact is initially quite weak, it significantly rises with the wavelength.

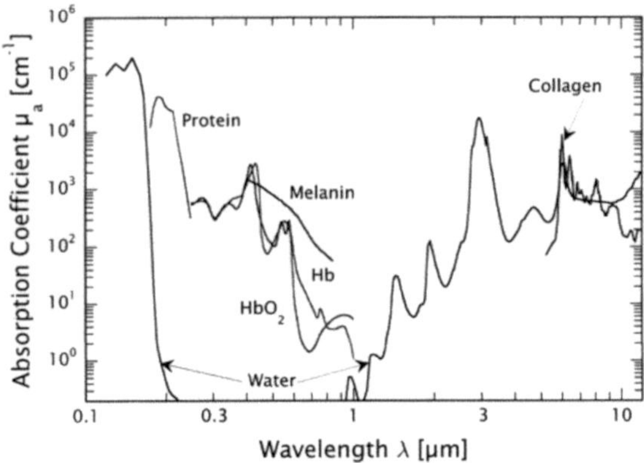

Figure 2.14: Optical absorption coefficients of main tissue components. (Reprinted with permission from [Vog03]. Copyright 2003 American Chemical Society.)

Scattering Properties

Optical scattering arises from spatial vibrations in the refractive index within the tissue and depends on the composition, size and morphology of cellular and extracellular tissue components [Wan00]. Significant scattering arises when there is a substantial variation in the refractive index over length scales, that are larger than or comparable to half of the wavelength of the incident light.

The optical penetration depth δ of the incident radiation is given by the reciprocal of the absorption coefficient and defines the depth to which the tissue is heated. However, when optical scattering is significant for a certain wavelength, δ is smaller than the reciprocal of the absorption coefficient and also depends on the diameter of the laser beam [Jac93].

Dynamics of Optical Tissue Properties

Whenever optical tissue properties are measured, this is usually performed under physiological conditions. However, pulsed laser ablation substantially changes the physiological conditions which in turn may lead to significant alterations of the optical properties. For example thermal denaturation may reduce the scattering coefficient. Changes in the tissue absorption coefficient also cause changes in the ablation behavior. Certain publications prove that the absorption peak of water ($\lambda = 2.94\,\mu m$) shifts towards shorter wavelengths for increasing temperatures. Furthermore a reduction of the absorption coefficient of tissue was reported under increasing temperature [Lan02].

2.6.3 Thermo-Mechanical Response

The volumetric energy distribution of pulsed laser irradiation causes significant thermal and mechanical transients, which are the driving force for all laser ablation processes that are not photochemically induced. The energy absorbed by the irradiated tissue is entirely converted into heat. The heated volume is typically a layer of thickness $1/\mu_\alpha$ and the thermal diffusion time t_d is given as

$$t_d = \frac{1}{\kappa \mu_\alpha^2},$$

with the thermal diffusivity κ and μ_α^2 the absorption coefficient. Once the energy is absorbed, spatial redistributions takes place by thermal diffusion [Wel95].

In order to achieve precise tissue ablation it is required to utilize a laser wavelength which causes only a small optical penetration depth and therefore a confined energy deposit within a small volume. Furthermore the spatial extent of thermal diffusion during irradiation needs to be limited in order to maximize the temperature in the absorbing volume. These thermal confinement is achieved if the ratio of laser pulse duration to thermal diffusion time is smaller or equal one.

2.6.4 Thermodynamics

Ablative cutting and material removal necessitates the fracture of chemical bonds in the irradiated tissue. Ablation is mostly driven by phase transitions from the solid, respectively liquid, to the vapor state.

2.6.5 Ablation Plume Dynamics for Short Pulses

Material removed from the ablation site forms an ablation plume. The flow of ablation products perpendicular to the tissue surface induce a recoil pressure that may cause additional material expulsion and collateral effects in the remaining tissue. Furthermore the deposition of laser radiation energy into the tissue is influenced by the ablation plume. Additionally the ablation plume dynamics also influences the interaction of the laser beam with the ejected material, since scattering and absorption of the incident light with the plume reduces the amount of energy deposited into the tissue and thereby limit the efficiency of ablation.

2.6.6 Ablation Models

The laser ablation process for biological tissue is complex and the variety of laser and tissue parameters influencing it is numerous. Therefore it is quite challenging to derive a single model that faithfully describes the physics of the process and predicts parameters such as the amount of tissue removal, the threshold radiant exposure for material removal and the zone of thermal injury.

Ablation Process Quantities

The ablation process is mostly characterized by quantities such as ablation threshold, ablation enthalpy and ablation efficiency. The ablation threshold Φ_{th} represents the minimum radiant exposure required to achieve effective laser ablation, i.e. volumetric material ejection. Pulsed laser ablation does occur with destruction of the ECM and not only by mere dehydration of the tissue.

In order to describe the energetics of the ablation process, the ablation enthalpy h_{abl} or ablation heat is applied [Hib97]. Conventionally the ablation enthalpy only considers the energy which is actually deposit into the tissue. However, it is crucial to take the absorption losses within the ablation plume, as well as the diffuse and specular reflection by the tissue into account. Mostly the radiant exposure is used, since it is difficult to determine the actual amount of energy deposited.

A metric for the total energy necessary to remove a defined mass of tissue is given by the ablation efficiency η_{abl}. The ablation efficiency is zero at the ablation threshold. For steady ablation, ablation increases monotonically and asymptotically reaches $1/h_{abl}$ for high radiant exposures. Considering blow-off ablation, the ablation efficiency reaches its maximum at moderate radiant exposures and decreases with higher radiant exposures. This describes the effect of energy wasted at higher exposures by overheating the tissue layers without contributing to the ablation process.

Heuristic Models

Early models for predicting the amount of material removed by laser ablation were heuristic ones without any consideration of the particular ablation mechanism. In fact these models are based on specific laser parameters and rely on empirically determined metrics as described in the previous section.

The *Blow-off Model* was developed in order to predict the material removal for ultra violet (UV) laser ablation of polymers [Deu83]. The characteristics of the model are illustrated in Figure 2.15 and are based on the following assumptions:

- The spatial distribution of absorbed laser energy is described by Beer-Lambert law,

- below the threshold radiant exposure the tissue is only heated and ablation is initiated by exceeding the threshold,

- material removal starts after the end of laser irradiation and

- the conditions for thermal confinement are fulfilled.

These assumptions hold for laser ablation with pulse durations below 100 ns. Then the depth δ of the crater resulting from laser ablation is in semilogarithmic relation with the incident radiant exposure and can be expressed as

$$\delta = \frac{1}{\mu_\alpha} \ln\left(\frac{\Phi_0}{\Phi_{th}}\right).$$

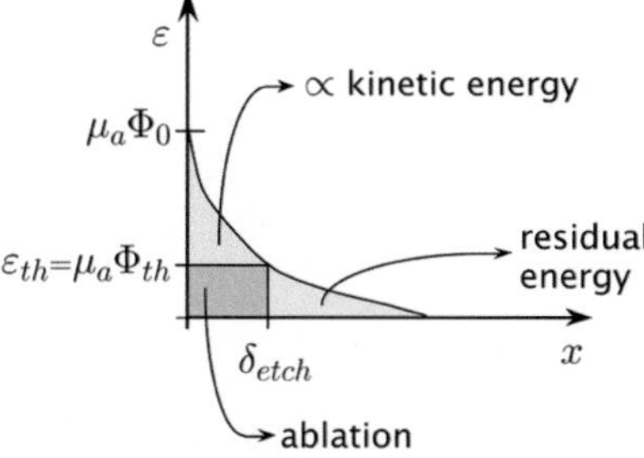

Figure 2.15: Scheme of the Blow-off Model. (Reprinted with permission from [Vog03]. Copyright 2003 American Chemical Society.)

Applying a laser with a wavelength at which the absorption coefficient of the tissue is large therefore results in a low threshold radiant exposure.

While the Blow-off Model is based on the assumptions that the ablation is initiated after the laser irradiation ends, the *Steady-State Model* takes into account that for laser pulses with microsecond duration (or longer) the material removal typically occurs concurrently with the irradiation. The continuous ablation process can be expressed under the assumption that a fixed energy density h_{abl} is necessary to remove a unit mass of tissue with density ρ. This leads to a rate of material removal which balances the irradiated energy to the tissue. In contrast to the Blow-off Model the material removal begins directly after the threshold radiant exposure is reached. The Steady-State Model then predicts the depth of the ablation crater in linear dependence to the incident light exposure

$$\delta = \frac{\Phi_0 - \Phi_{th}}{\rho h_{abl}}.$$

In this case, no direct dependency on the tissues absorption coefficient is made. In fact the threshold radiant exposure includes the absorption coefficient indirectly

$$\Phi_{th} = \frac{\rho h_{abl}}{\mu_\alpha}.$$

Mechanical Models

Since for heuristic models the mechanisms underlying the ablation process are not considered, developed mechanical models try to explicitly correlate laser and tissue parameters with the outcome of the ablation process, i.e. removed material and residual thermal injury.

Inspired by applications for metal ablation, first *Steady-State Vaporization Models* consider the ablation process as a rapid vaporization and assume that the boundary of the irradiated material moves. Even though the assumption of a fixed vaporization temperature does not meet thermodynamic principles, the moving boundary description of the ablation process is sufficient to model various aspects of the ablation process. The three-zone model introduced by McKenzie [McK86, McK90] was the first comprehensive model specifically for tissue ablation. This model covers the effect of the laser parameters to the excision and the thermal impact, i.e. the temperature profile inside the tissue and the thermal injury. The model does not include the thermal diffusion after finishing the laser irradiation, which also influences the thermal injury zone. Therefore Venugopalan et al. [Ven94] developed a model which allows to determine the zone of thermal injury and regards the thermal diffusion of the heat induced by the laser pulse.

Models which also take mechanical effects during the ablation into account are called *thermo-mechanical models* and were also based on models generated for metal ablation. Main contributions to thermo-mechanical modeling of the ablation process were made by Zweig and Frenz [Zwe87, Zwe88]. Here the tissue is considered to pass into a liquid phase first and then pass into a vapor phase. The transition from solid into liquid occurs upon thermal denaturation. The thermodynamics of the vaporization process are considered, but phase explosions are not taken into account. Material or liquid ejection does only occur as a byproduct of the surface vaporization recoil forces.

2.7 Medical Applications Using Laser

Lasers have become an integral part in medical applications these days. As short historical overview of some landmark publications of laser in medicine is given:

- Ophthalmology – Zaret et al. [Zar61]

- Dentistry – Goldman et al. [Gol64a]

- Dermatology – Goldman et al. [Gol64b]

- Laser bone interaction – Lithwick et al. [Lit64]

It is important to notice that the variability of laser parameters on the one hand offers numerous application areas, but on the other hand necessitates careful use. The same effect, which might be positive for a certain application could cause harm for another. The following subsections describe important aspects and applications of hard tissue ablation as well as the thermal effects and healing observations of hard tissue after laser treatment. Soft tissue laser processing is not within the scope of this doctoral thesis and therefore only literature for further reading is given [Ber03, Nie07]. Some general aspects are discussed in Section 2.5ff.

Generally laser provides several advantages, which make this energy source favorable especially in medical applications:

- Absence of vibration mechanisms and pressure,

- Removal of tissue difficult to access with conventional methods,

- Processing tissue precise and selective.

2.7.1 Hard Tissue Ablation Using Laser

Bone tissue is a composite structure which makes it an unfavorable material for laser processing: a high mechanical strength, relatively high thermal conductivity and a small amount of water content in conjunction with high melting temperature of its mineral components prevents the correct modeling and prediction of the ablation process itself.

Bony tissue is mainly composed of 67% inorganic minerals (hydroxyapatite) and 33% collagen and non-collagenous proteins. Water and hydroxyapatite are both high absorbers of laser light in the infrared wavelengths area and show peak absorption coefficients at around $3\,\mu m$ and $10\,\mu m$ wavelength. By focussing highly absorbed radiation of a laser pulsewise onto bony tissue, its superficial layer is heated up immediately while inertia prevents thermal expansion. The pressure within the irradiated volume increases. Thereby the internal pressure rises and by surpassing the bone's strength leads to micro-explosions that promote thermal induced mechanical tissue ablation (thermo-mechanical process). The overheated water in the tissue is vaporized and solid tissue fragments are carried away with the vapour during the micro-explosions [Web04]. The largest amount of the deposited heat is removed by the vapour and only a small amount diffuses into the adjacent tissue.

Most utilized laser wavelengths for hard tissue ablation are in the mid infrared. On the one hand this is the CO_2 laser $(9.6/10.6\,\mu m)$ and on the other hand the Er:YAG/YSGG laser $(2.94/2.79\,\mu m)$. The interaction of infrared lasers with hard tissue is mainly driven by thermal processes. Furthermore UV excimer lasers (e.g. XeCl, ArF) which cause direct photoablation and ultra-short pulsed laser systems (ps/fs pulse lengths) are applied, where the wavelength is negligible since high irradiance causes plasma-induced ablation or photodisruption. In the scope of this doctoral thesis an IR laser was utilized, therefore the state of the art for thermo-mechanical hard tissue ablation with focus on CO_2 lasers is given in the following.

Er:YAG/YSGG Hard Tissue Ablation

Er:YAG/YSGG lasers transmitting at wavelengths of 2.94/2.79 µm show the absorption maximum in water. Therefore irradiation directly *boils* the water and causes explosive evaporation. Er:YAG/YSGG lasers allow only pulse repetition rates up of to 100 Hz only and in comparison to CO_2 lasers show poor beam quality. This prevents fast and high volume removal. Therefore this laser types are mainly utilized in medical application which require small bone removal. For example, Erbium lasers are mostly utilized ablation of dental enamel (e.g. [Bad06, Ish08, Iar08]). Experimental osteotomy with Er:YAG lasers are described [Stü07b, Stü09, Pap07].

The advantage of this laser type is, that the radiation can be transmitted through optical fibers and therefore surgical hand pieces are designed to be manually guided by the surgeon. The fiber tip is in direct contact with the tissue. The laser radiation is transmitted through a water-vapor channel which occurs around the fiber tip by the leading parts of the laser beam.

CO_2 Laser Ablation of Bone

Throughout the beginning of studies with CO_2 laser in order to ablate hard tissue, continuous wave or long pulsed (in the range of ms) irradiation was chosen. The intuitive selection of laser parameters and long pulse durations to deposit high intensities in a technical easy way, caused severe thermal damage to the adjacent tissue with charring and carbonization and finally led to delayed healing. These first attempts did not indicate the feasibility of CO_2 laser to perform bone cutting. [Gol70, Cla78]

Since the late 1980s several publications addressed CO_2 laser ablation of hard tissue again. First publications addressed the delayed healing problem due to carbonization, but also encouraged the use of CO_2 lasers for hard tissue processing in regard of the advantages [Jam86]. Different short pulsed laser systems were utilized (in µs range) since the beginning of the 1990ies and experimental and histological results verified, that pulse durations well below the thermal relaxation time of hard tissue components combined with the strongly absorbed wavelength of the CO_2 laser, allow hard tissue ablation with acceptable thermal damages [For93].

In order to minimize the thermal damage of adjacent tissue, the laser parameters have to be chosen appropriately. If the wavelength of the laser is strongly absorbed by the tissue and the pulse intensity is high, while the pulse duration is short, the energy is quickly deposited into the tissue. This avoids thermal dispersion before the ablation process starts [Fre03]. To prevent tissue dehydration an applied air-water spray additionally cools the tissue [Iva98b, Iva98a, Iva00a, Iva00b, Iva02, Iva05b, Iva06]. In addition to fine water spray fast scanning of the laser beam avoids cumulative thermal effects by multi-pass irradiation [Iva05a, Wer07b, Wer07a]. In-vivo experiments revealed that effective CO_2 laser osteotomy is feasible without any significant delay in healing. Werner showed that cuts with a depth of more than 15 mm are possible [Wer06]. However, cuts created by short-pulsed CO_2 laser ablation underlie limitations in the achievable incision depth. This is due to the fact, that the ablation debris is confined in the incision and leads to an increasing amount of energy loss with increasing depth. Furthermore, the Gaussian energy distribution of the laser beam lead to characterically wedge shaped incision profiles. This profile causes the heat diffusion to become faster which in turn leads to decreased ablation outcome. Ablation halts if the accumulated energy in the tissue is not sufficient anymore to promote the process [Wer07b].

2.7.2 Thermal Effects

In case of bone ablation temperature elevation above 44 °C have to be avoided in order to prevent necrosis [Eyr05]. Dry ablation leads to strong thermal damage. Inspired by investigations in material processing, the use of water layers enhances the ablation performance and lowers the damage threshold.

Frentzen et al. [Fre03] analyzed ex-vivo osteotomies of rib and cortical bone specimen from a pig. They utilized a CO_2 slab laser (Rofin Sinar, SCx20) with 10.6 µm wavelength, pulse durations of 80 µs with an energy of 46 mJ at a repetition rate of 100 Hz while applying air-water spray (1-2ml saline/min). In all samples a char-free ablation without cracking adjacent tissue could be observed. The extent of collateral damage could not be correlated to the depth of the incision. No necrosis could be observed. CO_2 laser irradiation resulted in a darker basophilic stained layer in the histology.

The influence of the water layer thickness on the ablation result using an Er:YAG laser was investigated by Mir et al. [Mir08]. Compared to air-water spray, the bubble formation and channel propagation in water leads to a more symmetric ablation process.

Investigations on hard tissue ablation under different liquid environments were carried out with a long-pulsed Er,Cr:YSGG laser by Kang et al. [Kan08] (cp. Figure 2.16). Comparing five succeeding 150 μs pulses applied onto fresh bovine tibia by using dry, distilled water layer, perfluorocarbon layer and water spray showed that the ablation was optimal when using water spray. Even though perfluorocarbon reduces the energy loss due to lower IR absorption and therefore the ablation volume was about 15% higher, thermal damage was caused (characteristically black carbonization of the crater). Using water spray to support the ablation process achieved clean cuts with smooth crater surfaces without thermal damage. Water inhibits the temperature rise and the rapid vaporization of interstitial water promotes additional mechanical impact on the crater wall. Additionally Kang et al. assume that the water flow additionally causes a cleaning effect by removing debris. Characteristically formed craters (narrow, sharp-cone) are explained by forward

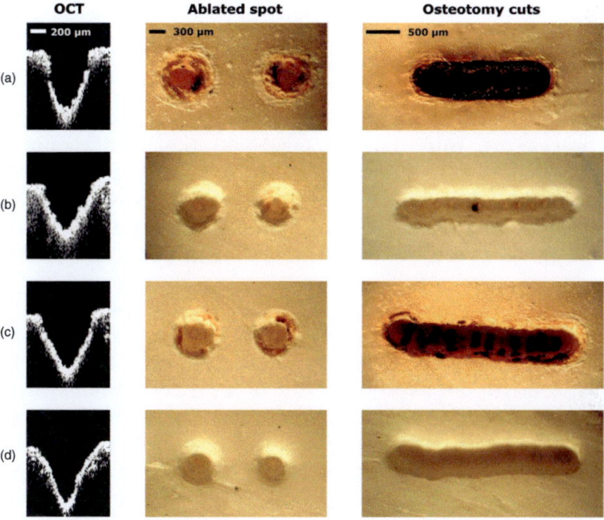

Figure 2.16: Results of the investigation performed by Kang et al. Cross-sectional OCT, top view onto ablated spots and onto osteotomy cuts under (a) dry, (b) water layer, (c) perfluorocarbon layer and (d) water spray ablation. (Reprinted with permission of [Kan08], Copyright 2008 IOP Publishing Ltd.)

scattering of the laser beam. Scattering contributes to decreased ablation volumes.

2.7.3 Healing

In a comparative histological study of Er:YAG laser (λ = 2.94 μm, focused, 10 Hz and 50 mJ per pulse) cuts with those achieved by a conventional bur (round steel, diameter 2 mm, 1500 rpm) on in vivo rat tibia de Mello et al. [deM08] showed that the bone repair is more advanced after 7 to 14 days for the laser cuts. Holes with a diameter and depth of 2 mm were cut for these experiments. After 21 days the histological results of both cuts were similar. This study also showed that the margin area of laser cut tissue showed characteristic layers which seems to promote short delays in the integration of new formed bone with the original bone.

2.8 Conclusion

The interaction mechanisms of laser with matter are the basis of laser material processing. Interaction of laser with biological tissue is mainly based on absorption. The main interaction types were summarized in this chapter. In case of laser bone processing thermal interaction is significant. The temperature rise inside the tissue promotes micro-explosive removal of tiny bone fragments and constitutes an ablation crater.

Understanding the mechanisms, that drive the ablation of biological tissue is essential. The comprehensive knowledge of physical, chemical and biological processes taking place under laser irradiation is of high importance towards an establishment of a process model for robot assisted laser osteotomy. Therefore an overview was provided within this chapter. Some details will be reflected in Chapter 4.

Several advantages of laser material processing make this source of energy favorable for industrial applications. Its improved quality and automation worthiness drove numerous applications, from which laser cutting and welding are most prominent. Since computer and robot assistance is utilized in both application areas, concepts and ideas can be accounted for realization of robot assisted laser processing of biological tissue. However, these methods can not be transferred or adapted offhand, since the focus for industrial applications differs significantly from the one for medical or surgical applications. This aspect is analyzed in detail in the next chapter.

Beside the utilization of laser in the industrial application area, it also gained importance in medicine. For example, laser in opthalmic surgery or dermatology fundamentally promoted medical care and opened up new medical procedures. The developments regarding hard tissue ablation were introduced in this chapter and expanded in the next chapter.

3

Computer and Robot Assisted Surgery

The emerging of computer science in the mid of last century had a profound if not massive impact on everyday life and revolutionized the handling of information. Even though surgeons performed surgery for thousands of years the gain in information thanks to computer science also influenced the state of the art in surgery in a fundamental way.

In this chapter computer assisted surgery will be reviewed and the main concepts will be introduced. Subsequently robot assisted surgery is discussed. Both sections provide a general overview of the broad field of computer and robot assisted surgery. Furthermore the state of the art of computer and robot assisted medical laser applications is introduced. The chapter ends with the open questions and requirements in order to develop robot assisted laser bone ablation.

3.1 Computer Assisted Surgery

Conventionally, the medical images preoperatively derived from the patient are put on a light box in the operation theater and the surgeon has to compose them spatially in his mind. Thereby he achieves a three-dimensional impression of the specific anatomy and pathology. A correspondence to the actual position of the patient at the operating table needs to relate the image information spatially to the patient.

43

The main achievement of computer assisted surgery (CAS) is the support of the surgeon in determining the spatial correspondence and in processing of information. On this basis, the support to the surgeon in localization and guidance is in the scope of computer assisted surgery. In the following the relevant research areas of computer assisted surgery are described and literature is referred for further reading.

3.1.1 Medical Imaging

Computer science with the possibility to compute large amounts of data in reasonable time changed medical imaging essentially. Computed tomography (CT) is one of the most important imaging modalities in the scope of CAS. Other modalities are conventional X-ray, magnetic resonance imaging (MRI), ultrasound (US), functional nuclear imaging as positron emission tomography (PET) or single photon emission computed tomography (SPECT) [Beu00].

Computed tomography utilizes X-ray in order to determine images of the target area. An X-ray source and a detector are interconnected and rotated around the target during scanning period. The data is progressively taken as the target is gradually passed through the gantry. This large amount of data has to be processed using tomographic reconstruction algorithms in order to derive 2d slices of the scanned volume. The comparison between exposed radiation intensity and measured intensity after traveling trough the target allows to relate to the composition of the scanned material. Named after the inventor of computed tomography Godfrey Hounsfield a quantitative measure of radiodensity is used for CT scans. The scale is defined in Hounsfield units (HU), running from air at -1000 HU, through water at 0 HU, and up to bone at +1000 HU. Major drawback of computed tomography is the comparatively high radiation needed for image acquisition. Newer approaches with flat panel detectors like cone beam CT (DVT) reduce the radiation dose.

3.1.2 Image Processing and Representation

After acquisition of medical images this data has to be processed in order to derive a model of the anatomy which is considered for the surgery. Such computational models can be distinguished according to the information reflected into: geometrical, physical and physiological [Del06]. In the scope of this doctoral thesis the physiological modeling of the patient is not of interest and will not be addressed. A comprehensive overview of image processing, representation and visualization can be found in the following literature [Fit00, Ban08, Pre07].

Geometrical Modeling

Geometrical modeling addresses construction of static description of the anatomy usually based on medical images. Patient specific models can be generated by various imaging modalities and may have different dimensions (two, three or three plus time). The extraction of geometric information from those images needs segmentation algorithms generating the targeted structure. Numerous techniques for segmentation of anatomical structures, either automatic or semi-automatic, have been developed during the past years.

The nature of the geometrical representation is strongly dependent on the application. For instance volume rendered patient information just requires labeling of the voxels representing a region of interest and definition of opacity functions. For computer assisted surgery usually a surface model derived from triangulation methods is determined. The transformation of segmented image data into a surface model requires the application of the marching cubes algorithm [Lor87]. Usually numerous triangles are created which tend to be poorly shaped. Therefore decimation techniques are utilized to reduce the number of triangles and to improve the shape. However, the main influencing parameter for the accuracy of the processed image data is the resolution of the imaging modality.

Physical Modeling

Physical modeling involves the behavior of various tissues. For instance, geometrical models only provide a static representation of the anatomy and do not take into account the deformation of soft tissue that may occur during surgery. Physical models address the issue of incorporating biomechanical information. In the scope of this doctoral thesis physical modeling also involves the volumetric removal of tissue (bone) under exposure of energy.

3.1.3 Planning

Planning in computer assisted surgery can either be performed pre-operatively or intraoperatively. The surgeon develops a surgical plan from a patient specific model or from a priori information about human anatomy contained in an anatomical atlas or database. Since surgical procedures differ significantly, planning is highly dependent on the surgical application. Some cases might involve a relatively straightforward interactive simulation or selection of the target position, such as performing a biopsy in neurosurgery. In other cases, such as in maxillofacial or orthopedic surgery, planning may require sophisticated optimizations incorporating tissue characteristics or biomechanical information adapted to the patient-specific model [Jos01].

3.1.4 Registration

In the scope of computer assisted surgery registration refers to the determination of the spatial correspondence between points represented in two different coordinate systems. Most prominent is registering the preoperatively acquired images to the patient on the operating table. The gold standard is point-based registration and will therefore be further explained in the following. Further registration methods, i.e. surface based registration or image to image registration, are stated in the referred literature [Mai98, Haj01].

Point-based Registration

This method takes the advantage of the approximate rigidity of the anatomy so that the registration can be accomplished by a well defined rigid-body transformation. Assuming a rigid transformation the registration leads to a bijective correlation between one coordinate system with any position vector $\vec{p}$ into another coordinate system with the corresponding position vector $\vec{\Phi}$. This mapping can be described as $\vec{\Phi}(\vec{p}) = R * \vec{p} + \vec{t}$ where R is a rotation matrix and $\vec{t}$ the translation vector. Let P be a point set (source points) and G be the point set (target points) to which the registration should be determined. Both point sets are of same size ($n \geq 3$). The correspondence is known ($p_i \in P \rightarrow g_i \in G$ for $i = 1..n$) and the points are described as position vectors ($\vec{p_i}, \vec{g_i}$). Then R and $\vec{t}$ can be determined by minimizing the sum of squares of residual error:

$$\varepsilon_{RMS} = \sqrt{\frac{1}{n} \sum_{i=1}^{n} (\vec{g_i} - \vec{\Phi}(\vec{p_i}))^2}.$$

A unique solution exists if and only if the point sets P and G contain at least three not collinear points. Several closed form solutions for the problem have been proposed, with the most prominent one presented by Horn [Hor87].

Implementation of Point-Based Registration in CAS

Usually fiducial markers (artificial landmarks, e.g. screws) are used to achieve the two corresponding point sets for registration. The fiducials are therefore implanted in a specific configuration in proximity to the area of surgical interest. After acquiring medical image data, the location of these fiducials is determined manual or (semi-)automatically. The locations of the fiducials in the image data are defined in the coordinate system of the images (also referred to as planning coordinate system). Intraoperatively, the medical image coordinate system needs to be spatially aligned with the patients' coordinate system. Hence, the locations of the fiducials are determined using a measuring device (e.g. localization system). After determining the point-based registration the correspondence between the planning and intraoperative coordinate system is known.

Point-based registration is the gold standard for computer assisted surgery. However, there are several limiting factors. First, it can be difficult to identify the fiducial markers precisely either in medical images as well as on the patient site. Second, the accuracy of point-based registration crucially depends on the number of fiducials and their configuration [Fit98].

Error Measures

Introduced by Fitzpatrick et al. were three commonly used measures of errors for analyzing the accuracy of point-based registration methods [Fit98]:

- **FLE**: The Fiducial Localization Error represents the error in locating the fiducial points.

- **FRE**: The Fiducial Registration Error is defined as the root mean square distance between corresponding fiducial points after applying registration.

- **TRE**: The Target Registration Error represents the distance between corresponding points other than the fiducial points after applying registration.

Fitzpatrick et al. derived a closed-form solution to estimate the expected squared value of the TRE in rigid-body point-based registration [Fit98]. Assuming an identical and isotropic Gaussian distribution of the FLE, the TRE depends crucially on the FLE, the number of fiducials and the fiducial configuration.

Usually the FRE is chosen to indicate the quality of a registration. Intuitively the surgeon relies on this measure and if the FRE is smaller, assumes that also the TRE must be smaller. However, Fitzpatrick and other researchers recently showed that this assumption in fact is wrong: FRE and TRE are uncorrelated [Fit09, Sha09]. They concluded that the FRE is a reasonable indicator of whether a system is functioning properly for a given procedure on a given patient or not, but, if the system is functioning properly, then no measure of the goodness of fiducial fit provides any information about the level of accuracy achieved by the system for that patient.

3.1.5 Navigated Surgery

In general, navigated surgery refers to the determination of the location of surgical instruments in the operation theater and the visualization of the spatial correspondence to medical images or a patient specific model. Hence, registration is a key issue for the accuracy of such systems. Tracking of instruments can be achieved by different measurement technologies, most prominent optical tracking. Furthermore electromagnetic tracking is a widely used method and the interested reader is referred to the following literature [Yan06, Mas05, Woo05, Yan09]. In order to achieve the most technical accurate registration, measurement arms are utilized which show high localization accuracy. Some aspects on navigated surgery will be described in detail in the following.

Optical Tracking

Optical tracking is based on triangulation. Two or more cameras are spatially aligned and observe the target area. In order to localize a surgical instrument or the patient, fiducials are attached, which can be detected in the camera images. A configuration of these artificial landmarks for optical tracking is also referred to as rigid-body or dynamic reference frame (DRF). The geometrical configuration of the rigid-bodies are matched with the measured values and with this information the transformation of the actual position of the rigid-body is determined.

Fiducials can be infrared light emitting diodes, reflective markers that are illuminated with infrared light by the tracking system or markers that exhibit high contrast in the visible spectrum. Optical tracking systems provide accuracy in sub millimeter range, refresh rates that are sufficient for the most medical procedures and robust performance with regard to the environment [Kha00]. However, a drawback of optical tracking is the necessity of line-of-sight with the tracked instruments.

Surgical Guidance

Supporting the surgeon in spatially aligning the medical images or pre-operative plan with the actual situation in the operating theater is the major goal of surgical guidance. After successful registration of the patient, the medical images and the intraoperative location of the patient are aligned. By tracking a pointer or a surgical instrument the location within the medical images can be determined and visualized for the surgeon. By defining the surgical target, e.g. a tumor or the placement of a biopsy needle in neurosurgery, the surgeon can now navigate his instrument within the images. Figure 3.1 illustrates a state of the art navigation system.

Beyond simply navigating in the medical images, surgical guidance can also be implemented by additional support. For instance, by providing distance information to the target either numerically or visually by geometric primitives and colors.

Navigated Control

Beyond surgical guidance navigated surgery also includes navigated control, where the surgeon is actively supported in performing a sur-

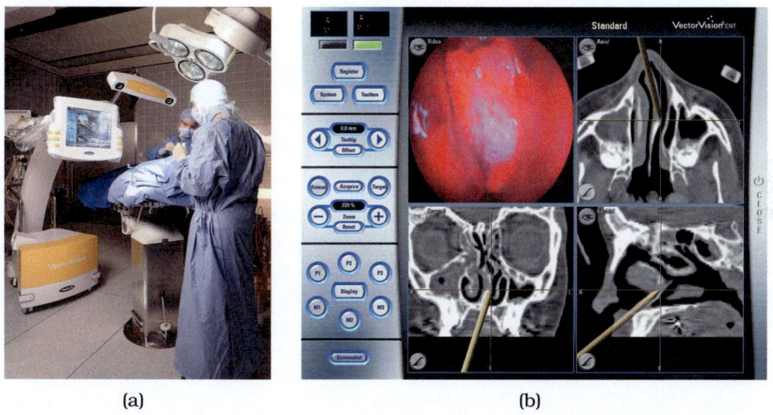

(a)   (b)

Figure 3.1: (a) Typical surgical navigation system in the operation theater. (b) The location of the instrument is visualized in the medical images. (Pictures: BrainLAB AG)

gical task. The navigation information is utilized in order to control a surgical burr or drill. By defining the surgical target in the medical image data, for example a volume to be removed, risk structures and forbidden areas, the location of the instrument in the medical images can be interpreted. Thereby the instrument is either switched on or off, or controlled in power. Navigated control was introduced by Lüth et al. [Szy03] and applied to numerous surgical applications [Hof08].

3.2 Surgical Robotics

Medical robotics is an interdisciplinary research field which is principally application-driven. Especially in this research field fundamental advances in technology must provide significant and measurable advantages for being acceptable by the medical community.

In comparison to the industrial application field, the emphasis in surgical robotics is not put on the automation idea, which massively changed productivity and efficiency. In fact the robot is regarded as a surgical tool, which is enhancing the abilities of a surgeon and not replacing him. A significant improvement of the surgeons' technical capabilities by a surgical robot can be achieved by enhancing an existing procedure in increasing the accuracy, lowering the time needed or invasiveness. Furthermore surgical robotics can facilitate otherwise infeasible interventions. In general one can say, that the potential of surgical robotics emerges from the complementary strengths of humans and robotic devices, as summarized e.g. by Taylor and Joskowicz [Tay03b] and given in Table 3.1.

3.2.1 Milestones of Surgical Robotics

The first medical application of a robot was performed in 1985 in stereotactic neurosurgery by Kwoh et al. [Kwo88], where the robot was used to position a device for orienting a needle for brain biopsy. The biopsy target was identified in CT images of the patient and a PUMA 560 robot was utilized to orient the guide tube through which the surgeon inserted the needle. This first application of a robot to surgery served as a prototype for the commercialized system Neuromate from Integrated Surgical Systems. Neuromate gained FDA (U.S. Food and Drug Administration) approval in 1999 for passive use in image-guided stereotactic brain surgery under control of the surgeon. Integrated Surgical Systems Inc. was one of the earliest companies in the field of surgical robotics, founded in 1990 in Sacramento, California. It has developed

Table 3.1: Complementary strength and limitation of humans and robots: (++) excellent, (+) good, (-) limited, (--) hard. (Adapted from [Tay03b])

Ability	Human	Robot
Integrate and act on multiple information sources	++	-
Versatile and able to improvise	++	--
Geometric accuracy	+	++
Dexterity	+	++
Judgment	++	--
Hand-eye coordination	++	-
Untiring and stable	-	+
Resistant to radiation and infection	--	++
Adapt to new situation	++	--
Integrate multiple sources of sensor data	-	++
Sterility	-	++

two image-directed, semi-autonomous robotic products for neurological and orthopedic surgical applications.

In an initial euphoria, partially autonomous surgical robotic systems for total hip or knee replacement [Tay94b] were commercialized very soon. The ROBODOC system was applied clinically in 1992 for the femoral implant component [Sch07]. Originally the ROBODOC system was developed as a prototype at IBM research during the late 1980s and further developed clinically by Integrated Surgical Systems. 1994 commercialization began in Europe. Until 2004 ROBODOC was criticized harshly regarding the technique, results and complications caused by the system and the clinical use was mostly discontinued except for India, South Korea and Japan [Sch07]. The ROBODOC Surgical Assistance System has been cleared recently by the FDA for Total Hip Arthroplasty procedures as reported by the ROBODOC company (division of CUREXO Technology Corporation, Sacramento, USA). Today ROBODOC is the only active robotic system cleared by the FDA for orthopedic surgery. The system has been used in over 24000 surgical procedures assisting surgeons around the world. However, no significant advantage of the robotic assistance could be reported [Sch07].

Around the mid 1980s surgical robotics research began in Asia [Ise94] and Europe [Ben87, Dav91]. From then on, surgical robotics research has been applied to almost every clinical application area (neurosurgery, orthopedics, urology, cardiology). Most of the systems remained in the research or pre-clinical evaluation phase and only a few system gained access to clinical routine. For further reading the interested reader should refer to the specific publications [Fau06, Dav06, Cle01, Cle05a, Cle06, Tay03a, Tay06, Sic08, Hoc07].

3.2.2 Interaction Categories

The main advantage of applying a robot in a surgical intervention is seen in assisting the surgeon either in performing a surgical subtask (semi-)autonomous or in direct interaction. Hence, one fundamental categorization is made in terms of the level of autonomy of a surgical robot:

Autonomous

Surgical robotic systems belonging to this category perform a specific task autonomously. Usually a preoperatively planned procedure is carried out by the robot and surveyed by the surgeon. Applications are characterized by the requirement of complex and repetitive optimal paths, which would be impossible for a human to follow with sufficient accuracy. The acceptance of applications out of this category is rather difficult, because of the historical complications (e.g. ROBODOC, PROBOT). An autonomous robotic systems needs to be highly reliable. However, from the research point of view, this interaction category poses the most challenging questions and offers promising medical applications, which may not be feasible to be manually performed by a surgeon.

Interactive

Interactive or semi-automatic surgical robotic systems are characterized by the shared control between surgeon and robot. The surgeon supervises and controls a robot, while he is actively constrained by the robot. This approach increases safety regarding patient-robot interactions and promises higher accuracy. The surgeon remaining in control of the procedure improves the acceptance of these applications.

Telemanipulated

Surgical robotic systems in this category are explicitly controlled by the surgeon. The motion performed by the surgeon on a master console is transferred to the robot(s). This implies physically separateness of the surgeon from the surgical robot and therefore is also referred to as tele-operation, even though the surgeon is usually located in the operating room with the robotic system. Nowadays the acceptance of telemanipulated robotic systems is the highest for surgical robotics.

The first idea behind telemanipulated surgical robots was to allow a surgeon to perform an operation from a considerable distance and thereby avoid exposition to hazardous conditions in wartime or after natural disasters [Hil98]. The central problem of telemanipulated robotic systems is the communication delay.

3.2.3 Workflow of Autonomous Robot Assisted Interventions

Surgical robotics for automated or semi-automated execution of a defined surgical task, incorporates a preoperative phase, where the intervention is planned computer-assisted. The intraoperative phase needs registration of the patient in respect to the robot and updating of the preoperative plan to the intraoperative situation. Eventually the patient model is updated using intraoperative medical imaging and the plan is updated correspondingly. The robot assisted execution is mostly related to a specific part of the overall surgical intervention, e.g. removing a volume of bone for preparing an implantation.

3.2.4 Mechanical Design

The mechanical design of a surgical robot is strongly dependent on the intended application. However, most applications were realized with more or less essentially modified general purpose industrial robots. Offering many advantages, especially in research, like low cost and high reliability [Cas00], these industrial systems come along with high speed and also large footprint. While industrial applications aim at maximizing the performance and the processing speed, exact implementation is the main focus in surgical interventions. But the specific requirements of surgical applications generally imply specialized robots. The emerging field of minimally invasive surgery for example induced the development of specialized surgical robots providing high degrees of dexterity in constrained spaces inside the patient, including the most successful commercial telemanipulation system daVinci. The daVinci surgical system is a laparoscopic robotic system from Intuitive Surgical Inc. (Sunnyvale, USA). However, the first commercial laparoscopic system for minimally invasive surgery was the ZEUS Robotic Surgical System from Computer Motion, Inc. which was approved in Europe, but FDA clearance was missing. In 2003, Intuitive Surgical and Computer Motion agreed to merge.

The patient as an inexact described *workpiece* with his individual anatomy and pathology necessitates *one of a kind work*. Even though surgeons fixate or immobilize parts of the patient during surgery, the in-vivo surgical target to be treated with a robotic system necessitates continuous supervision. Methods of navigated surgery are therefore combined with robotic systems. Furthermore additional sensory is a key issue for safe surgical robotic systems (e.g. force sensing).

3.2.5 Bone Treating Robotic Applications

Thanks to its rigidity, bone is a favorable target for surgical robot applications. Hence, many surgical robotic systems deal with the treatment of bone and most of them are autonomous or interactive systems. In contrast, soft tissue applications are subject of telemanipulated surgical robotics.

First application of a robot in surgical treatment of bone was already in the beginning of surgical robotics research in orthopedics for preparing a total hip or knee replacement (cp. Section 3.2.1). Robotic systems for similar applications are in the scope of research since then on. Furthermore, there were also commercialized systems as CASPAR (ortoMaquet, Rastatt, Germany) [Mei98] and Acrobot (The Acrobot Company Ltd, London, UK) [Dav06]. However, also these system did not gain success and vanished from the market. Nowadays, these kind of systems are emerging the market again, e.g. MAKO Surgical Systems. Still utilizing a robot, the systems are now operated by the surgeon and in the category of interactive/semi-autonomous surgical robotics.

A further application field for surgical robotics is around the skull, hence in neurosurgery, cranio-maxillofacial surgery and ear-, nose- and throat-surgery (ENT). The first complex trajectory on a patient was autonomously performed in the scope of the collaborative research center 414 of the German research foundation [Eng03, Kor03, Wör06]. A modified Stäubli RX90B CR robot was utilized in order to perform a preoperatively planned cutting trajectory on a human skull for craniotomy (cp. Figure 3.2). Numerous applications of surgical robotics in research can be found in the literature, but none of them gained access to the clinical practice.

Another bone treating application for surgical robots is spine surgery. Here, the aim is to provide support in accurate positioning of implants to the surgeons(e.g. pedicle screws) and reduce the intraoperative radi-

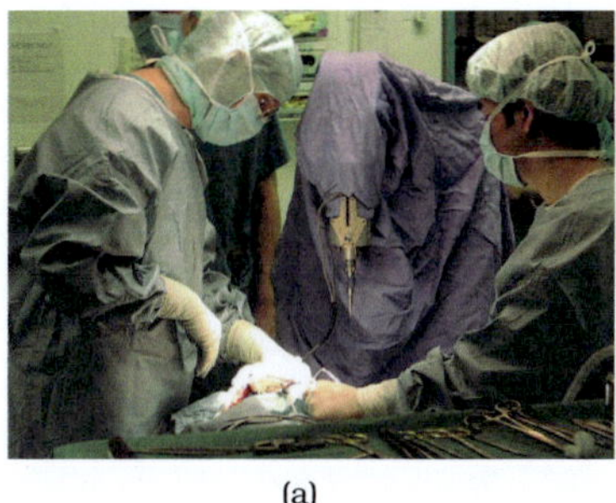
(a)

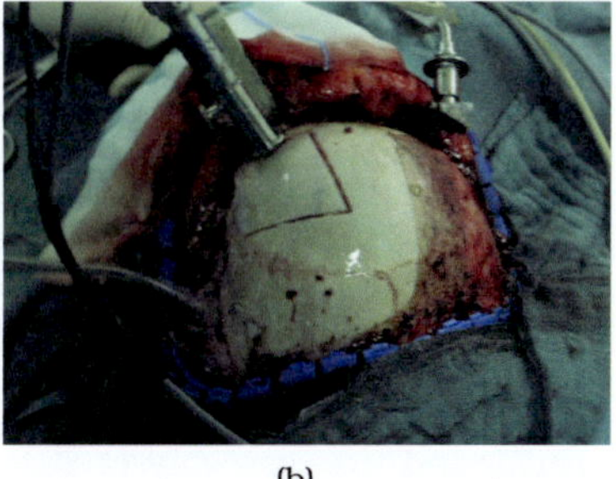
(b)

Figure 3.2: (a) Setup of the RobaCKa surgical robot in the operation theater. (b) Robot-assisted craniotomy on a patient skull. [Eng03]

ation exposure. For these applications parallel robots are often utilized, e.g. the commercial system *SpineAssist* (Mazor Surgical Technologies, Israel).

Nowadays application of robotics around cochlear implantation is discussed controversially and several applications are proposed. For example Federspil and Stolka et al. use a robot for preparing the skull bone for the implant [Sto07a, Fed03] and several research groups are utilizing robotic assistance for other tasks connected with cochlear implantation, i.e. drilling a hole into the lateral skull for inserting the electrode [Kle09, Maj09, Bre07, Cou08]. Further applications of surgical robotics in the cranial area can be found in [Wid07, Bas06].

3.2.6 Current Developments

As already pointed out, research in surgical robotics is application driven. Developed systems are usually constrained to a specific intervention. Nowadays the research efforts emerge towards more versatile robot systems, which are no longer capable of performing just one intervention type. The main design questions for versatility focus around catchwords as *lighter* and *smaller* in order to facilitate easy integration into the operation theater [Hag08, Sho03, Pec09]. Furthermore the trend towards performing more and more surgeries minimally invasive is amplifying the development of miniaturized surgical tools with rich sensory in order to give the surgeon back missing sensation, which isthe major drawback of laparoscopic surgery.

Among the development of versatile surgical robot systems, the commercialization is a key issue. About 25 years after evolvement of the research area surgical robotics only a few systems are commercially available and used in clinical practice. On the one hand this is due to necessary technical improvements [Mei98, Kno07, Cru07] and on the other hand due to the rather restrained enthusiasm of industry and surgeons. Important points towards bringing autonomous or interactive robotic systems into the operation theater are among legal questions (liability) and safety issues. Nevertheless, the growth of the surgical robotics industry will be amplified by the fact of our rapidly aging society and the necessity of best care and quality in surgery [Wan06]. However, a surgical robot must provide significant and measurable advantages for being acceptable in the medical community [Tay06].

As stated in the report about the operating room of the future standards for devices and their use are solely needed [Cle05b]. Furthermore interoperability of devices is essential, since stand-alone systems do not reflect the technological possibilities as it is state of the art in other areas (*plug and play*).

3.3 Computer and Robot Assisted Medical Laser Applications

While most industrial laser applications utilize computer and robot assistance (cp. Section 2.4), laser radiation in medical applications is mostly conducted manually by hand. However, some research work is performed in this scope. In the following publications will be introduced that are somehow connected to the topic of this doctoral thesis. This part concludes with the summary of the preliminary work performed at the Institute for Process Control and Robotics (IPR) of the Universität Karlsruhe (TH).

3.3.1 Conceptual Work

Buzug et al. proposed a navigation concept for image-guided laser surgery in order to improve conventional surgical implantation procedures. For gaining a better localization accuracy, the authors suggested a holographic technique. Furthermore *laser drilling* with a CO_2 laser was performed and preliminary tests in bony specimen using a special beam scanning procedure were executed but not explained any further. For controlling the process they suggested to measure the acoustic signal with a piezoelectric transducer. They state that in homogeneous tissue a precision of up to $\pm 100\,\mu m$ in determining the ablation depth is achievable. Additionally they analyzed the acoustic spectra in order to differentiate tissue, but even though they found material dependent features they did not state any results. To conclude the work presented by Buzug et al. they proposed several methods for laser surgery but without claiming how to connect them for a complete system.

Rizun et al. aimed in developing a tactile feedback laser system for robotic surgery and proposed a conceptual design and presented the first prototype [Riz05], which allows haptic feedback in the vertical direction and keeps the orientation of the laser fixed. The manipulator was not evaluated in any experiment and it remains unclear how to apply the concept to an application. Proof-of-concept was stated in the publication, but not shown.

In the scope of laparoscopic surgery, there are some concepts to introduce laser as a tool for treating soft-tissue. Smith presented his idea for combining optical fiber diode lasers for cardiac surgery with the daVinci surgical robot system [Smi06]. Even though he had no concept in mind, Smith expects that coupling the laser methodology and surgical robotics will allow for more precise direction of the laser energy. Furthermore he anticipates that robotic dissection has got the potential to add a level of safety to cardiac interventions by enhancing visualization and control. A further proof-of-concept study was presented by the K.U.Leuven. The researchers designed and developed a dedicated robot system *VESALSIUS* for the purpose of CO_2 laser laparoscopy [Tan05].

3.3.2 Urology

Boris et al. applied a KTP laser as cutting and hemostatic tool for partial laparoscopic nephrectomy [Bor07]. The green light KTP laser with a wavelength of $532\,nm$ was delivered through a flexible glass fiber to a micro wrist instrument of the daVinci surgical robot system (Intuitive Surgical, Sunnyvale, USA). The ports were manually set and the surgeon telemanipulated the daVinci for this application. The potential steadiness of the laser held by a robot facilitates cutting and coagulation at much lower power than achievable with hand held lasers, which underlie unavoidable tremor, which causes in return beam scattering. The authors conclude that with the hand tremor reduced application of the laser, it will become a more precise and effective tool in future. However, the conclusion was also that the present state of the KTP laser is ineffective for cutting, but feasible for hemostasis.

3.3.3 Orthopedics

Kim et al. propose a surgical robot for the rotational acetabular osteotomy mounted on iliac bone to perform the incision using an Er:YAG laser irradiating [Kim04]. A passive holder with detachment mechanism is meant to be directly mounted to the cortical bone. Attached will be a parallel link mechanisms which holds the laser handpiece. The authors utilize an Er:YAG laser with wavelength of 2.94 µm, operated with a repetition rate of 20 Hz and a pulse duration of 200 ms. Beam delivery is realized by an optical glass fluoride fiber with a sapphire contact tip. The distance between the bone and the fiber tip was kept at 0.25 mm with a force sensor and spring. The authors published the evaluation of the mechanism and the structure of the manipulator. The manipulator itself is not specified any further and the adjustment to the bone is performed manually. However, the work does not comprise any planning or registration procedure.

3.3.4 Oral and Cranio-Maxillofacial Surgery

Rupprecht et al. presented Er:YAG laser osteotomy controlled by a sensor system [Rup03, Rup04]. Figure 3.3 illustrates the control loop. They utilize on the one hand a piezoelectric accelerometer mounted directly on the bone surface (contact area 50 mm^2) in order to detect vibrations and a silicon photodiode to analyze the luminous phenomena (photosensitivity 350-1100 nm). The feedback system switches off the laser when exceeding threshold values which are found to be characteristic for the transition from cortical to cancellous bone. The system was evaluated for rabbit femur first [Rup03] and then for in vitro mini pig mandibles for ten drilled holes [Rup04]. It was possible to determine high frequencies above 200 kHz during the processing which are explained with the explosive vaporization. The frequencies giving a hint about the transition between cortical and cancellous bone are significantly lower (1-10 kHz). The authors concluded that it is possible to detect acoustic signals, but this is hard due to much static and background noise. Therefore an exact feedback system could not be realized. Furthermore the authors state that sensitive structures could be preserved with the feedback system. Their results were proven radiographically and histologically on dissected mini pig jaws.

Stübinger et al. utilized the Er:YAG laser for extraction of a displaced tooth in one patient case [Stü07a]. The laser with a pulse energy of 500 mJ, 250 µs pulse duration and a repetition rate of 12 Hz was guided through an optical fiber with a diameter of 1000 µm. Stübinger et

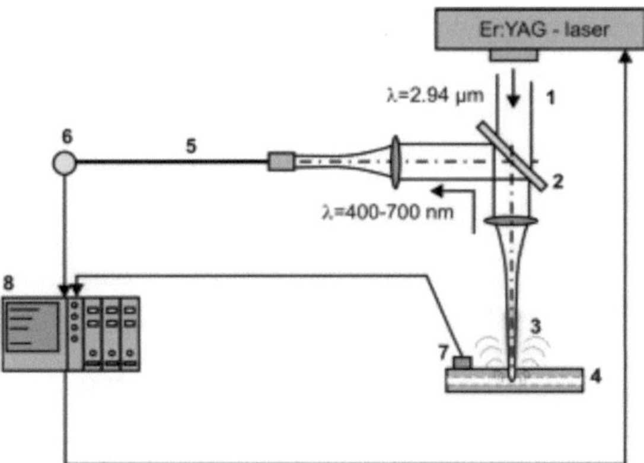

Figure 3.3: Circuit of closed-loop control system 1 = Er:YAG laser beam; 2 = beam divider; 3 = process emissions; 4 = bone specimen; 5 = optical fiber; 6 = photodiode; 7 = piezoelectric accelerometer; 8 = digital signal processor. (Reprinted from [Rup03], Copyright 2003, with permission from Elsevier.)

al. stated that a distance of 1-2 mm was kept, but not how this was achieved and one may reasonably expect that this was done manually by the surgeon. However, the intervention was preoperatively planned by the analysis of the 3d reconstructed patient model and a so called *dynamical simulation* of two access variants, which is not described in detail. To summarize the computer assistance in that application is characterized by the preoperative 3d reconstruction of the patients anatomy and the visualization. The surgeon plans the intervention in his mind, supported by additonal information (e.g. bone density on the basis of the Hounsfield values). Hence, the laser process itself is not supported by computer assistance, neither during planning nor during the application. The authors note that the osteotomy gap is limited by the diameter of the fiber but without stating any achieved numbers. Therefore it is questionable that the stated minimized bone loss compared to conventional osteotomy tools was achieved.

Stopp et al. recently presented a system for navigated laser surgery for dental implantology [Sto07b, Sto08c, Sto08b, Hoh09]. Motivated by the advantages of laser bone processing compared to conventional cutting methods and the lack of precise navigation, they proposed a system which utilizes a unfocused Er:YAG laser navigated and power-controlled.

In order to support the surgeon in manually performing laser ablation of bone with a handpiece, an optical navigation system is continually tracking the hand piece and the mandible. The current location of the hand-piece is visualized in presurgically acquired CT data and a three-dimensional simulation of the material removal. The system prevents the surgeon to ablate bone outside the cavity and to use the laser out of the destinated distance. Figure 3.3.4 shows the principal setup of the system.

This approach is characterized by the first application of navigated control methods to laser bone ablation and minimal changes in the intraoperative workflow. Navigated control allows to preserve risk structures. However, the presented methods, especially for the simulation of the bone removal are currently lacking of a well understood and mathematically described ablation process. Furthermore the manual use of a laser hand-piece does not allow precise bone ablation in the micrometer range. In this context the achievable accuracy of navigation systems (around 1 mm) is also considerably limiting the overall accuracy.

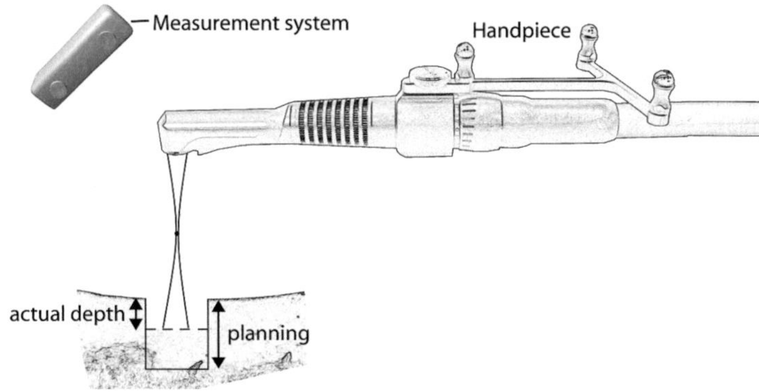

Figure 3.4: (a) Volume model of the energy distribution resulting from irradiation. (b) Setup for navigated laser surgery. (Reprinted from [Sto08a], Copyright 2008, with permission from VDI Verlag GmbH.

3.3.5 ENT Surgery

Solares and Desai et al. presented a system for transoral robot assisted CO_2 supraglottic laryngectomy [Sol07, Des08]. They equipped the daVinci surgical robot with a hollow core fiber (OmniGuide Inc., Cambridge Massachusetts, USA) and performed cadaver and canine experiments, before applying the methods to patient trials. In nine patient cases they showed feasibility of their methods using a continuous wave CO_2 laser at 10-14 W for treating tumors of the upper aerodigestive tract. To conclude the authors presented a system for telemanipulation utilizing fiber guided CO_2 laser radiation.

3.3.6 Ophthalmic Surgery

LASIK (laser in situ keratomileusis) surgery with its good results in visual perception enhancement has become a widely available and accepted procedure to correct refractive errors. The concept of customized ablation was developed in the mid 1990s and is based on corneal topography. The corneal tissue removed during cornea reshaping for a specific amount of correction is less as compared to the conventional method. Stojanovic et al. presented in their work alternatives for customized ablation planning for corneal refractive surgery, especially irregular astigmatism [Sto05]. In order to optimize the ablation the authors utilized the software CIPTA (corneal interactive programmed topographic ablation) which uses corneal anterior surface elevation maps, measured by a topographer. On this basis the software explores ablation possibilities by considering the corneal surface from purely morphological aspects and is therefore uncoupled from the optical axis. The authors concluded that the system provides excellent results for virgin eyes in general. To conclude, the customized ablation for cornea reshaping is quite advanced and used in the clinical practice. However, the developed methods are strongly restricted to eye surgery.

3.3.7 Laser Osteotomy

Kuttenberger et al. applied an early version of the CO_2 laser osteotome developed by the former group for Holography and Laser Technology at the Center of Advanced European Studies and Research (caesar, Bonn, Germany) for cutting sheep tibia [Kut08]. In two groups bone healing was compared between cuts achieved by the laser osteotome and a conventional oscillating saw. The laser ablation was performed in three stages using 80 µs pulses at 200 Hz and a resulting pulse energy of 75-85 mJ which were scanned corresponding to a manually set ablation pattern with 40 mm/s, whereby in the first third 1.2 mm widening of the cut was used using the wobble-scan technique. The second third of the osteotomy was performed reducing the widening to 0.6 mm and the last third without any widening. The focus was manually adjusted between the three stages. After osteotomy, osteosynthesis of the bone was conducted. The laser cut bone showed the disadvantage that the wedge-shaped osteotomy does not allow close contact between the two bone pieces. No visible signs of carbonization could be detected macroscopically and no fundamental differences between healing for laser or saw group could be determined. However, the authors stated that the vertical changing of the focus by moving the scanner along its optical axis was critical and therefore propose to include the laser into a robot guided and computer assisted navigation system. It was concluded that the missing tactile feedback during laser osteotomy is disadvantageous. The mechanical achieved cuts took considerably shorter than laser cuts. Figure 3.5 illustrates the experimental setup for osteotomy

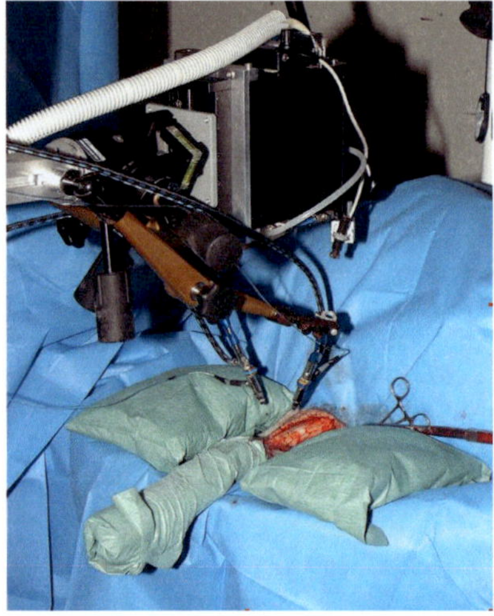

Figure 3.5: Experimental setup for osteotomy of an exposed sheep tibia using the CO_2 laser osteotome. (With kind permission from Springer Science+Business Media: [Kut10])

of the sheep tibia. One can conclude, that this work showed feasibility of cutting bone using laser, but lacks in methods for registration and positioning in submillimeter range. Therefore required accuracy cannot be assured.

3.3.8 Preliminary Work at the Own Institute

In the scope of the Collaborative Research Center 414 of the German Research Foundation a preliminary setup trial for robot assisted laser bone ablation was accomplished already in 2005 [Pet05, Wör05]. This experiment was performed in strong collaboration with the Institute for Laser Medicine (ILM, Heinrich-Heine University, Düsseldorf, Germany) and the former group for Holography and Laser Technology at the Center of Advanced European Studies and Research (caesar, Bonn, Germany). Figure 3.6 illustrates the experimental setup. The relatively cumbersome and heavy scan head was mounted to a Stäubli RX90B CR. An ex-vivo cadaver pig was manually situated in the focal distance to the scan head. The robot remains in its' position during the ablation

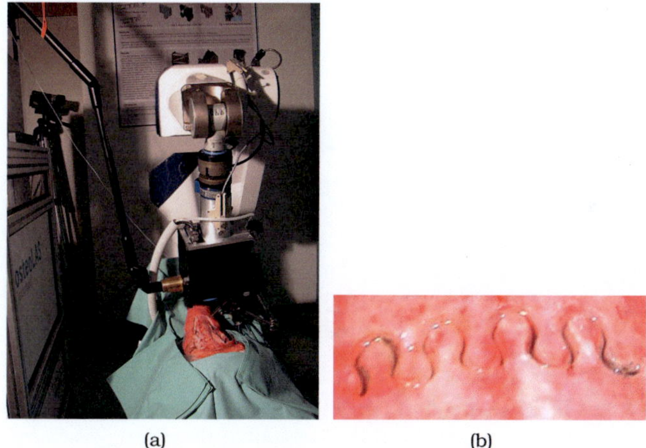

(a)   (b)

Figure 3.6: Preliminary trial for robot assisted laser bone ablation in 2005. (a) A cadaver pig was manually arranged under the scan head which was attached to a Stäubli RX90B CR robot. (b) A meander-shaped ablation path was performed.

of a meander-shaped ablation pattern. However, the experiment did not include any registration of the pig or computer assisted planning procedure. The ablation pattern was manually set with the scan heads' controller software. The cut was not completed and no osteotomy was performed. However, the principal feasibility of a robotic setup was demonstrated with this initial trial.

In the scope of the Priority Programme 1124 of the German Research Foundation visual controlled laser based cochleostomy was realized by Kahrs et al. [Kah07, Kah08, Kah09a]. The aim was to preserve the inner lining membrane of the cochlear until the insertion of the cochlear electrode (cp. Figure 3.7a). Kahrs et al. developed methods for visual control

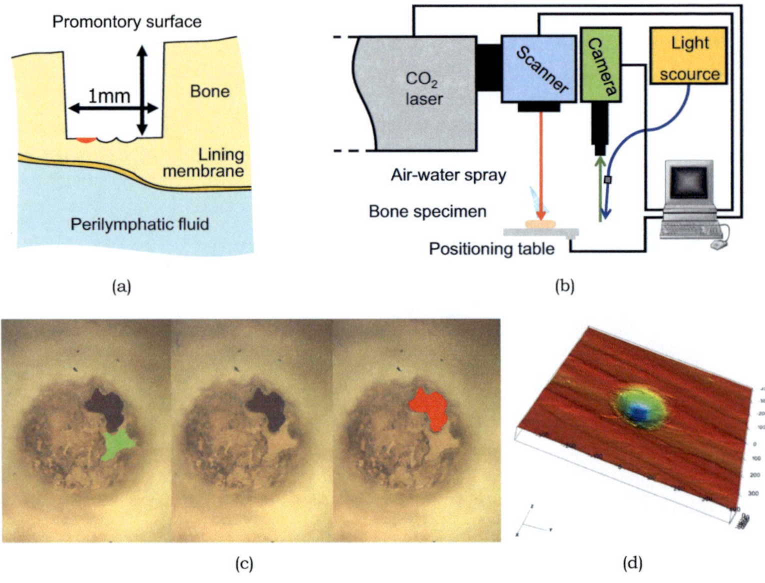

Figure 3.7: Visual controlled laser based cochleostomy. (a) Illustration of the anatomy and the need of application of several pulses for cochleostomy. (b) System setup. (c) Automatic segmentation of the membrane (green) and breakthrough (red) in the images for visual control. (d) Confocal microscopic measurement of a single laser pulse applied onto bone. (Courtesy of Kahrs.)

of the ablation process using a video camera with microscopic lens. In order to perform the cochleostomy the laser ablation is performed layer by layer and under surveillance (cp. Figure 3.7b). Image processing and automatic segmentation of the membrane (or breakthrough respectively) allowed for the first time microsurgical bone removal using laser for performing a cochleostomy (cp. Figure 3.7c). Furthermore Kahrs et al. performed planning and simulation of the bone removal on single laser pulse level [Kah09b] and analyzed for the first time single laser pulse induced ablation craters under a confocal microscope for quantitative evaluation (cp. Figure 3.7d). To conclude, Kahrs performed laser based drilling, but no cutting was performed. Furthermore his work does not include registration.

3.4 Conclusion and Open Questions

This doctoral thesis aims in answering the main question, whether robot assisted laser osteotomy is feasible or not. First of all it has to be pointed out, that nowadays robot assisted laser bone ablation for osteotomy is not existent. To describe the state of the art in the related research areas, Chapter 2 and the previous sections gave an overview of actual applications in CAS and developments in laser material processing.

At the first glance industrial laser applications utilizing robots and robot assisted laser osteotomy seem to have many things in common. However, methods used in industrial applications for robot assisted processing of material using laser are not adaptable offhand to medical applications. The patient as an inexact described *workpiece* with his individual anatomy and pathology ultimately necessitates *one of a kind work*. Furthermore, different requirements characterize these two application fields. Emphasis in surgical robotics is not put on the automation idea, increasing of productivity or efficiency. Nevertheless, industrial robotic laser applications provide hints and ideas which build a basis for medical applications.

The computer and robot assisted medical laser applications introduced in Section 3.3 are somehow related to the topic of this doctoral thesis. However, one can summarize, that each application offers a certain characteristic which is of relevance for robot assisted laser osteotomy, but no research work offers all characteristics at once. Table 3.2 shows the assessment.

Beside the question of the general feasibility of robot assisted laser osteotomy, this doctoral thesis also aims in providing answers to the following questions, which are unacknowledged in the state of the art nowadays:

- How to realize laser cutting of bone with robotic methods?

- Which cutting accuracies are required and obtainable with a system for robot assisted laser osteotomy?

- How to plan hard tissue ablation preoperatively?

- Which parameters are essential for controlling and optimizing the ablation process?

- Which surgical disciplines profit from robot assisted laser osteotomy?

Table 3.2: Assessment of characteristics for current computer and robot assisted laser applications: (++) excellent, (+) good, (-) limited, (--) nonexistent or (o) unknown.

Characteristic	Industrial Chapter 2	Boris [Bor07]	Kim [Kim04]	Stübinger [Stü07a]	Stopp [Sto08c]	Solares [Sol07]	LASIK [Sto05]	Kuttenberger [Kut08]	Kahrs [Kah08]
Laser material processing	++	+	+	+	+	+	+	+	+
Process model for bone removal	--	-	--	--	-	--	+	-	+
Complex trajectories	++	-	-	-	o	-	-	-	--
Cutting width down to 200 µm	++	o	o	--	-	o	+	+	+
Robot assistance (automatic positioning)	++	--	--	o	--	+	--	--	-
Overall accuracy below 0.5 mm	++	-	o	o	-	o	o	o	o
Computer assisted planning (patient specific)	--	--	--	-	+	--	++	--	+
Process accompanied by simulation	-	--	--	-	+	--	+	--	+

4

Modeling the Discrete Ablation Process

Dealing with the question how to implement robot assisted laser osteotomy leads to a first question: How does laser bone ablation work? Chapter 2 introduced the main aspects for processing of biological tissue using laser. In this chapter the fundamentals of short-pulsed CO_2 laser ablation will be reconsidered and preliminary work will be explained more detailed. The emphasis of this chapter is laid on modeling the discrete ablation process.

Initially the experimentation methods utilized in the scope of the laser ablation process analysis are introduced. Specific parameters of the utilized laser system are stated in order to facilitate comparability. After describing the scanning system for applying laser pulses onto bone, the optical path is characterized as an important property of the system.

Laser bone ablation is a process with multiple influencing parameters. With respect to the aimed computer and robotic assistance, the most important process parameters are identified and described. These are: Inclination angle, focal distance, depth development, catalyzing fluids and thermal dispersion. The influence of these parameters on the ablation process is evaluated.

In a further step, the process of bone removal caused by single laser pulses and their concatenation is presented. Hence, the process of bone cutting as consolidation of optical scanning and single laser pulse concatenation is explained. The chapter is concluded by a summary of the unique findings for the discrete ablation process model.

4.1 Fundamentals

From the laser material processing point of view, bony tissue is an unfavorable material: high mechanical strength, relatively high thermal conductivity and small water content in addition to a high melting temperature of its mineral components. Pulsed laser ablation is a complex process dependent on the properties of the irradiated biological tissue and with multiple influencing parameters. This prevents the correct modeling and prediction of the ablation process itself.

Laser ablation is mainly driven by the spatial distribution of the irradiated volumetric energy density. The volumetric energy distribution causes significant thermal and mechanical transients - the driving force for material removal. The energy absorbed by the irradiated tissue is entirely converted into heat. The heated volume is typically a layer of thickness, which corresponds approximately with the reciprocal value of the absorption coefficient $(1/\mu_\alpha)$. In order to achieve precise tissue ablation it is required to utilize a laser wavelength which causes only a small optical penetration depth and therefore a confined energy deposit in a small volume.

The wavelength of the CO_2 laser of $10.6\,\mu m$ is strongly absorbed by the two main components of bony tissue, i.e. hydroxiapatide and water. With a high pulse intensity and short pulse duration, the energy is deposited quickly into the tissue. Thereby thermal dispersion is avoided before the ablation process starts. By focusing this highly absorbed radiation onto the bony tissue, its superficial layer is heated up immediately, while inertia prevents thermal expansion. The pressure within the irradiated volume increases. Thereby the internal pressure rises and by surpassing the bone's strength leads to micro explosions that promote thermal induced mechanical tissue ablation (thermo-mechanical process). The overheated water in the tissue is vaporized and solid tissue fragments are carried away with the vapor during the micro explosions. The largest amount of the deposit heat is removed by the vapor and only a small amount diffuses into the adjacent tissue. To prevent tissue dehydration an air-water spray is applied, which additionally cools the tissue. In addition to fine water spray fast scanning of the laser beam avoids cumulative thermal effects by multi-pass irradiation [Iva05a].

Even though the state of the art reveals physical, chemical and biological dependencies and interrelationships, cutting bony tissue by the means of laser was regarded on the one hand theoretically and on the other hand phenomenologically until now. One has to notice that nowadays no comprehensive understanding of laser bone ablation and all influencing parameters is existent. Furthermore pulsed laser ablation for cutting bony tissue as a discrete process per definition was not considered on the basis of single laser pulses and their impact on bone removal until now.

4.2 Preliminary Work

In the scope of the doctoral thesis from Kahrs [Kah09a] single pulse based laser ablation for cochleostomy was performed. Therefore single laser pulses have been analyzed three-dimensionally using a confocal microscope. A two-dimensional Gaussian function fitting was performed semi-automatically in order to model the bone removal per pulse. For the first time Kahrs determined experimentally the parameters for modeling the bony volume removed by a single laser pulse. The results were used to automatically plan an optimal laser pulse pattern in order to ablate the cochleostomy channel. Figure 4.1 illustrates the preplanned simulated result of ablating three successive layer of pulses (right) and the measured result of applying the pulses onto bony specimen (left). Kahrs et al. showed that microsurgical bone removal using laser ablation is feasible for preoperative planning [Kah09b].

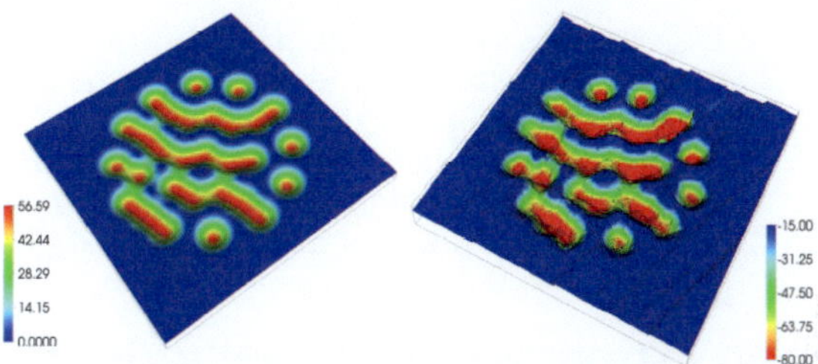

Figure 4.1: Simulation (left) and confocal measurement (right) of several single laser pulses. The examined area was $1.6\text{x}1.6\,\text{mm}^2$ and the values of the colored depth are in µm [Kah09b].

Furthermore he analyzed experimentally the correlation between pulse length and maximal depth of single laser pulses applied onto bony specimen. Two focal lenses with different focal depth were utilized for the experiment. The results are given in Figure 4.2. Due to the small amount of measured pulses, the variations are quite large. Amongst others Kahrs observed a linear dependency of pulse length and ablation depth.

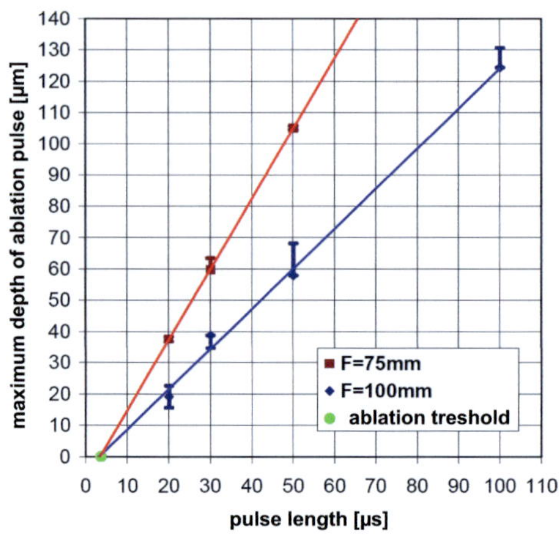

Figure 4.2: Maximum ablation pulse depth plotted against the pulse length for two flat field lenses with different focal distance. With courtesy of Kahrs.

However, the experiments carried out by Kahrs are the first step towards a model of the CO_2 laser bone ablation process. The discretization performed in the scope of this doctoral thesis is motivated by the work of Kahrs and will take more parameters into account. Furthermore the ablation crater analysis performed by Kahrs et al. [Kah09a, Kah09b] is investigated with a much larger data base.

4.3 Specific Experimentation Methods and Parameters

Even though the laser bone ablation process comprises complex physical and chemical interactions the result of a single laser pulse applied is reproducible and corresponds to a crater formed by several micro explosions. From the geometrical point of view one can say, that several of these single laser pulses applied result in the concatenation of craters. This section introduces the utilized bony specimen preparation and measurement methods. Furthermore the modeling of ablation craters as 2d Gaussian functions is described and the conclusions presented in the scope of cutting techniques are summarized.

4.3.1 Measurements of Removed Bone

Understanding the dependencies of laser parameters and the resulting ablation craters necessitates measurement of a sufficient amount of specimen. In the scope of this doctoral thesis several laser pulses in their concatenation have been evaluated. As bony specimen fresh frozen ex-vivo femoral cow bone was used. The bone was cut into slices using a diamond coated band saw in order to prevent metal abrasion, which could falsify the ablation results. The following two measurement modalities were applied.

Pulse Measurements

For precise quantitative measurement of laser pulses a confocal microscope (µSurf explorer, NanoFocus AG, Oberhausen, Germany) is utilized. Figure 4.3 illustrates the microscope and exemplary measurement data. In the scope of this doctoral thesis a measurement field of 800x800 µm^2 with resolution in x- and y-direction of 1.6 µm and resolution in z-direction of 6 nm was utilized. The microscope allows measurement of ablation craters with a maximum depth of 1000 µm. The resulting 3d topometry data is used for further analysis.

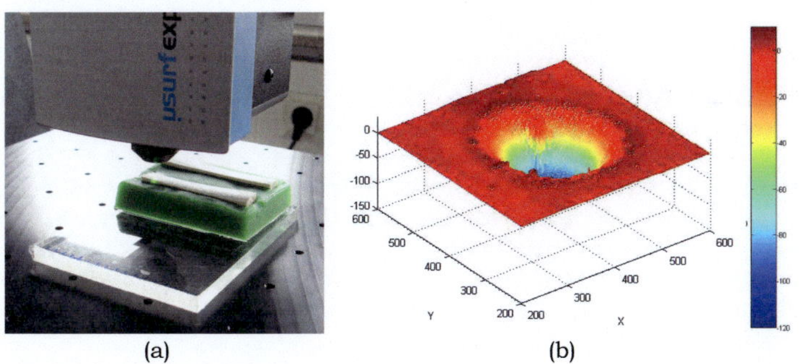

Figure 4.3: Confocal microscope µSurf (NanoFocus AG, Oberhausen, Germany) measuring the surface of bony specimen (a) and detailed visualization of the measurement (b).

Cut Measurements

For qualitative assessment of cuts deeper than the confocally measurable 1 mm a digital microscope (VHX-600, Keyence Corp., Osaka, Japan) is utilized. In comparison with a conventional optical microscope the digital microscope provides a depth of field at least 20 times larger. Figure 4.4 illustrates the microscope and an exemplary picture taken from bony specimen.

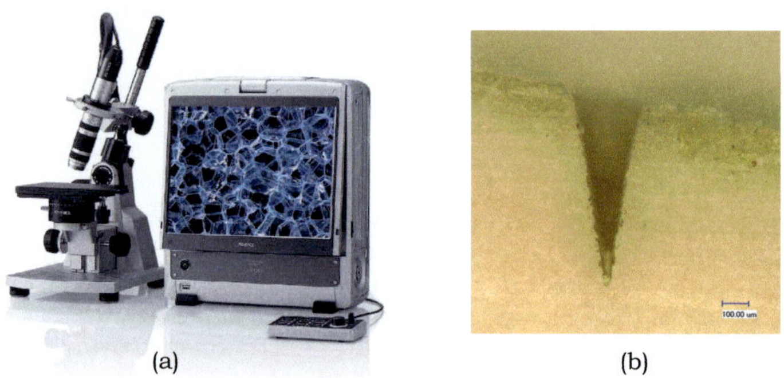

(a)   (b)

Figure 4.4: Digital microscope (VHX-600, Keyence Corp., Osaka, Japan) and an exemplary microscopic view on a cutting profile of bony specimen.

4.3.2 Data Analysis for Parametric Description of Ablation Craters

In the scope of this thesis 211 single laser pulses were applied onto bony specimen (femoral cow bone) and the resulting crater were measured using the confocal microscope. The measurements (matrix with values for the measured depth) were automatically analyzed with Mat-Lab [Pla09, Zha09]. In the following the preprocessing and 2d Gaussian function fitting will be described.

A single laser pulse produces a crater corresponding to the Gaussian profile of the laser beam. Hence, the crater can be described with a two-dimensional Gaussian function, which is defined as

$$f(x,y) = A \cdot \exp\left(-\frac{1}{2\left(1-r^2\right)} \cdot \left[\frac{(x-x_0)^2}{\sigma_1^2} - \frac{2r\left(x-x_0\right)\left(y-y_0\right)}{\sigma_1\sigma_2} + \frac{(y-y_0)^2}{\sigma_2^2}\right]\right)$$

with $A, x_0, y_0, |r| < 1, \sigma_1 > 0, \sigma_2 > 0$ are constant. Figure 4.5 illustrates a two-dimensional Gaussian function with the peak center at (x_0, y_0) where the amplitude A is reached. σ_1 and σ_2 are the standard deviation of the x- and y- marginal distribution and thereby define the spread of the Gaussian bell. The correlation coefficient r indicates the rotation of the Gaussian bell.

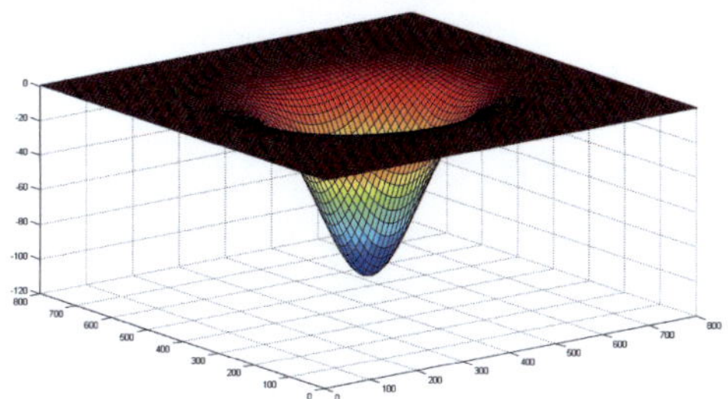

Figure 4.5: 2d Gaussian function with $x_0 = y_0 = 400, \sigma_1 = 128, \sigma_2 = 108, r = 0, A = -103$.

Preprocessing

The measured data of the confocal microscope shows noise as well as data points where no reflection could be measured (cp. Figue 4.6 left), especially when the measured crater is near the maximum of measurable depth. The preprocessing step therefore starts with noise reduction by applying a Gaussian low pass filter onto Fast-Fourier transformed data.

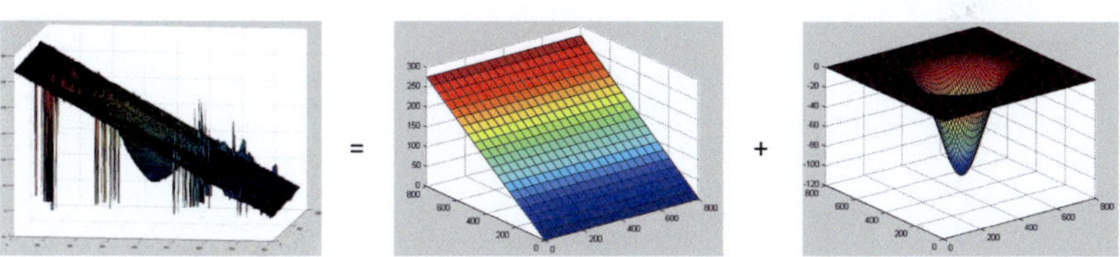

Figure 4.6: With a laser beam inclination angle of $\neq 0°$ the resulting geometry of the ablation crater is supposed to correspond to the conjunction of the 2d Gaussian function in direction of the beam with surface.

Since bony specimen is an unfavorable material in the sense that it cannot be absolutely flat, the laser beam impinges the bone with some inclination angle. In order to achieve comparable results it is furthermore necessary to align the measured data. With a laser beam inclination angle of $\neq 0°$ the resulting geometry of the ablation crater is supposed to correspond to the conjunction of the 2d Gaussian function in direction of the beam with surface (cp. Figure 4.6). Therefore the bone surface is fitted with a plane and from that the raw data is subtracted afterwards.

2d Gaussian Function Fitting

After preprocessing a dataset it is automatically fitted by a 2d Gaussian function. In order to not determine local minimum only, an appropriate starting value of the fitting vector $[x_0, y_0, \sigma_1, \sigma_2, r, A]$ is of high importance. By empirical observation, the minimum of the measurement matrix mostly occurs in the crater and corresponds to the maximum depth. Therefore the start value of the amplitude A is set to this minimum value z_{min}. The center of the Gaussian function (x_0, y_0) is then set to (x_{min}, y_{min}) of the minimum coordinates. The start value of r is set to 0, since x and y are considered to be independent from each other. σ_1 and σ_2 are both set to the $R/2$, where R is the empirical radius of the crater. The non-linear fitting problem is solved in the least squares sense (Curve Fitting Toolbox, MatLab). Figure 4.7 illustrates the confocal measurement of a single laser pulse and the corresponding 2d Gaussian function fitting.

After determining the 2d Gaussian function representing a specific ablation crater, it can be further analyzed. The following parameters are determined for evaluation:

- Radius of the crater

- Depth of the crater

- Volume of the crater

4.3.3 Energy per Pulse

In the scope of this doctoral thesis a short-pulsed CO_2 laser is utilized for bone ablation. The energy per pulse is an important parameter in order to give comparable findings. Using a power meter the mean power for pulses with durations τ from 20 to 100 µs with repetition rates of 200, 400 and 800 Hz was determined. Therefore an optical power

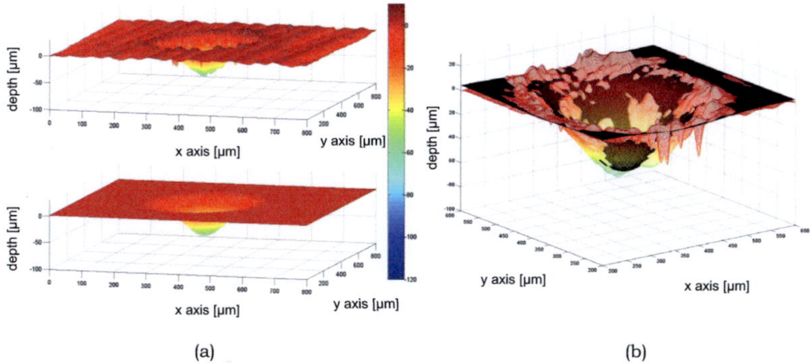

(a)   (b)

Figure 4.7: (a) Confocal measurement of a single laser pulse applied onto bone and corresponding 2d Gaussian function fitting. (b) Excerpt of a measured ablation crater overlaid with the corresponding 2d Gaussian function fitting.

meter (LPT-A-200-D25-HPB, Laser 2000 GmbH, Wessling, Germany) was used. Figure 4.8 illustrated the relation of mean and peak power for laser pulses.

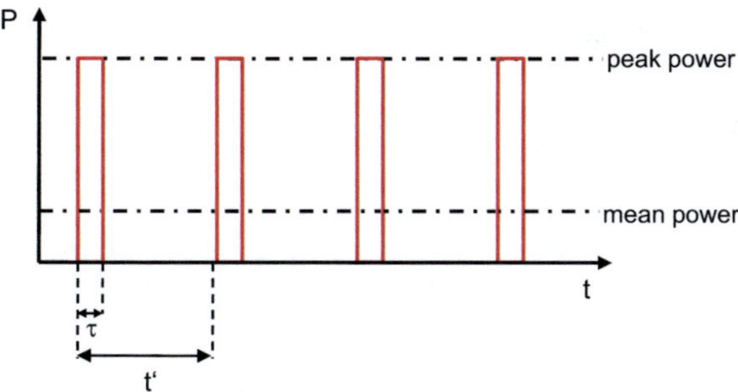

Figure 4.8: Relation of peak power to mean power. The pulse duration τ and the period time t' which is the reciprocal to the repetition rate.

From the mean power the energy per pulse can be directly derived. Figure 4.9 shows the results for the experimental laser system utilized in the scope of this thesis.

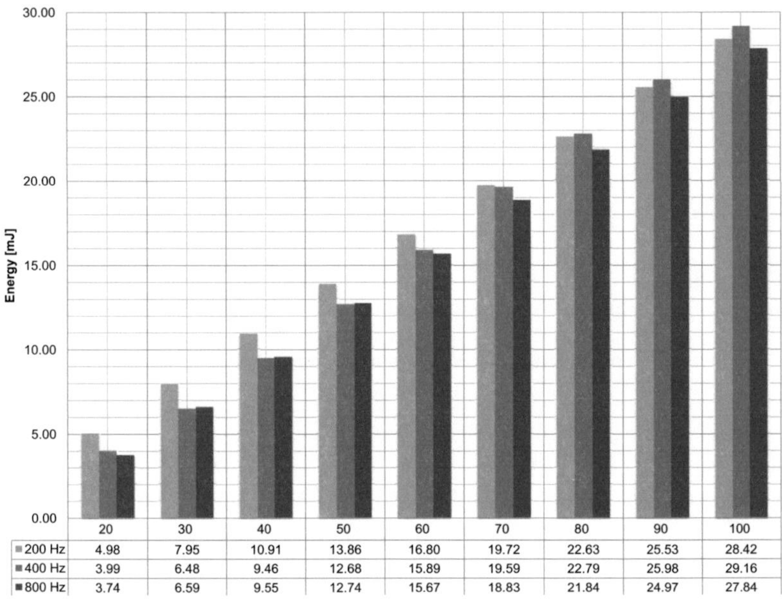

	20	30	40	50	60	70	80	90	100
■ 200 Hz	4.98	7.95	10.91	13.86	16.80	19.72	22.63	25.53	28.42
■ 400 Hz	3.99	6.48	9.46	12.68	15.89	19.59	22.79	25.98	29.16
■ 800 Hz	3.74	6.59	9.55	12.74	15.67	18.83	21.84	24.97	27.84

Figure 4.9: Energy per pulse corresponding to pulse duration (20-100 µs) and repetition rate (200, 400, 800 Hz).

Ablation Variance

Analyzing ablation craters revealed relatively high variances in the resulting ablation depth. Therefore the variance for 211 single laser pulses applied onto bony specimen were evaluated. The histogram in Figure 4.10 illustrates the variances expressed in percent. The variance is fittable with a Gaussian distribution. The maximum of the distribution is at 102% and with that slightly higher than the mean depth and most craters vary about ±5%. Considering the specified laser parameters of the utilized CO_2 slab laser which state a stability of ±7%, the results of this experiments become understandable. Furthermore the

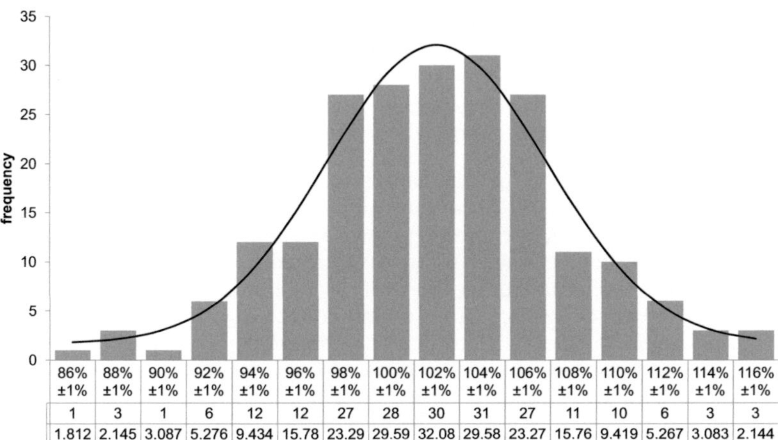

	86% ±1%	88% ±1%	90% ±1%	92% ±1%	94% ±1%	96% ±1%	98% ±1%	100% ±1%	102% ±1%	104% ±1%	106% ±1%	108% ±1%	110% ±1%	112% ±1%	114% ±1%	116% ±1%
	1	3	1	6	12	12	27	28	30	31	27	11	10	6	3	3
	1.812	2.145	3.087	5.276	9.434	15.78	23.29	29.59	32.08	29.58	23.27	15.76	9.419	5.267	3.083	2.144

Figure 4.10: Histogram for variance in the resulting crater depth of 211 single laser pulses applied onto bony specimen expressed in percent.

ablation is strongly dependent on the composition of the bone, basically i.e. water and mineral component content, density and spongiform.

4.3.4 Usage of a Scan Head

Laser ablation as a discrete process where single laser pulses are applied onto tissue in order to achieve removal of volume necessitates fast, repeatable and defined application of pulses. Hence, the method of choice is optical scanning, where mirrors are used to sweep the laser beam over an area. Generally there are three types of optical scanning configurations:

Objective Scanning The most common method is objective scanning, where the objective, laser source, image plane or a combination of these is moved [Mar04]. The scanning process is translational and relatively slow.

Post-objective Scanning In order to allow faster scanning, the deflectors are situated before or after the objective lens. Considering the objective lens as a reference position in the optical path through the scanning system, one then distinguishes between pre-objective and post-objective scanning [Bei03]. The simplest optical system is the post-objective scanning. In this configuration the rotation axis of the mirrors is mostly situated orthogonal to the optical axis. Focusing of the beam is realized prior to the scanning unit. The resulting scan field is typically curved.

Pre-objective Scanning In case of pre-objective scanning systems, the laser beam is scanned into an angular field first and then imaged onto a flat surface. Depending on the demand of the application for the optical correction required over a finite scan field, the scan lens is chosen which defines the spot size, scan linearity, astigmatism and depth of focus. Often multi-element flat-field (so called f-θ) lenses are utilized in pre-objective scanning systems. In the following only two-dimensional pre-objective scanning systems with a f-θ lens are considered when referring to scan heads.

In the scope of this thesis two-dimensional optical scanning is considered. Utilizing a two-dimensional pre-objective scan head for laser ablation necessitates understanding of the working principle as well as the parameters which describe the behavior when deflecting the laser beam. In the following the working volume of a scan head and scan field distortion are described.

Working Volume

The working volume of a two-dimensional scan head is per definition a plane if a flat-field lens is utilized for focusing. This flat plane is located in the focal distance of the lens. The x- and y-axes of the scan field are generated by the rotational axes of the two mirrors inside the scan head. In the scope of this thesis the z-axis of the scan field coordinate system is defined orthogonal to the x- and y-axis. A tolerance area around the scan field is defined which is corresponding to the Rayleigh distance of the focused laser beam. The resulting working volume of the scan head is therefore corresponding to the frustum of a pyramid. Figure 4.11 illustrates the working volume of a scan head.

Scan Field Distortion

The scan field underlies distortion comprising inherent, optical and mechanical factors. However, each optical system utilizing a lens underlies distortion and purely optical correction is limited. Correction by software is therefore indispensable. Usually the correction necessitates

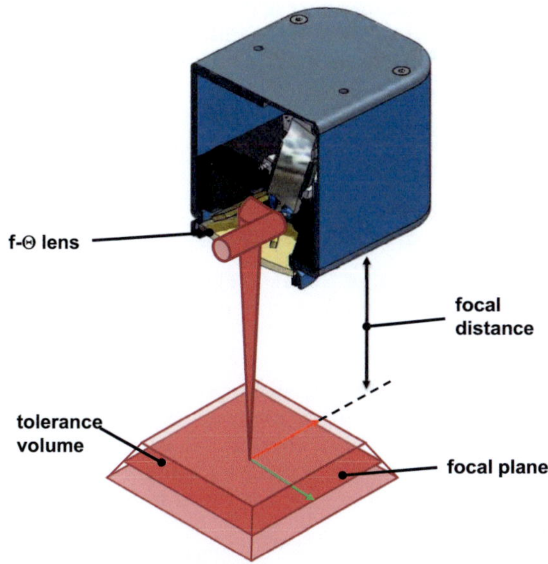

Figure 4.11: The scan head working volume is defined by the mounted f-θ lens, which focuses the coupled laser beam onto a plane in the focal distance. The tolerance volume is theoretically corresponding to the frustum of a pyramid. (Scan head drawing courtesy of Arges GmbH.)

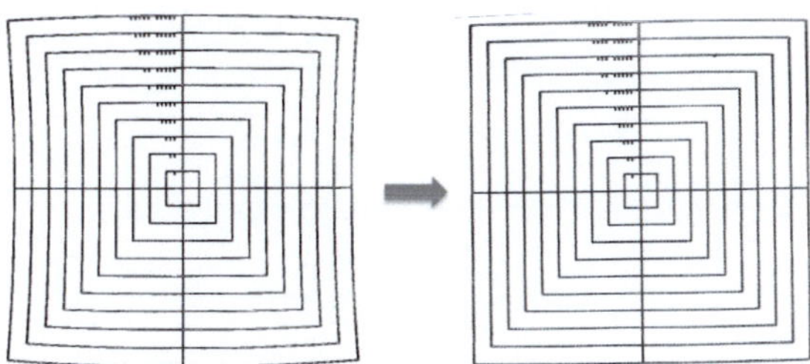

Figure 4.12: Without any correction the scan field shows a barrel distortion (left). Applying the correction method leads to an undistorted scan field (right).

scanning of a pattern and measuring of resulting distances, points, etc. An optimization algorithm then determines the correction parameters which lead to a corrected scan field with minimum distortion (see for example [Che07, Xie05]). Figure 4.12 illustrates the scan field before and after correction of the distortion. It is important to notice that determination of the right correction is correlated to a correct distance in which the correction pattern was applied to the target material.

4.3.5 Optical Path, Spot Size and Shape

The geometrical dimension of the focal spot is dependent on the length of the optical path and on the focusing optics or aperture utilized. The general dependencies in the following will be illustrated with specific values for the CO_2 laser (TEM$_{00}$, M^2 <1.2 and $\lambda = 10.6\,\mu m$). Leaving the laser resonator the laser beam has a specified diameter $D = 7.5\,mm$ and would normally show a diverging behavior. An optic at the output behind the beam forming unit already focuses the beam. The focal distance is $z_f = 1\,m$, where the beam has a diameter of $2w_{\sigma 0} = 7.2\,mm$. Hence, the Rayleigh distance z_{R0} of the beam waist $w_{\sigma 0}$ is given by

$$z_{R0} = \frac{\pi \cdot w_{\sigma 0}^2}{M^2 \cdot \lambda} > 3.2m.$$

After propagating additional 1.8 m (cp. Appendix B) from the beam waist, the laser beam is focused again (specified focal distance $z_f' = 101.3\,mm$). Figure 4.13 illustrates the optical path.

If the distance $z - z_f$ is much larger than z_f', the diameter $2w_{\sigma 0}'$ and the focus shift Δz of the second beam waist, are calculated following Beyer et al. [Bey98]:

$$w_{\sigma 0}' = \frac{w_{\sigma 0} \cdot f}{\sqrt{(z - z_f)^2 + z_{R0}^2}} \approx 99\mu m,$$

$$\Delta z = \frac{z - z_f \cdot f^2}{(z - z_f)^2 + z_{R0}^2} \approx 1.4mm.$$

The second Rayleigh distance can than be calculated:

$$z_{R0}' = \frac{\pi \cdot w_{\sigma 0}'^2}{M^2 \cdot \lambda} > 2.4mm. \tag{4.1}$$

The CO_2 laser in TEM$_{00}$ shows a Gaussian profile. The focal spot is defined to be circular in the focal plane. A beam quality factor $M^2 \neq 1$ modifies the spot from being circular to an elliptic shape. An additional effect on reshaping the beam profile is discussed in the following (Section 4.4.1).

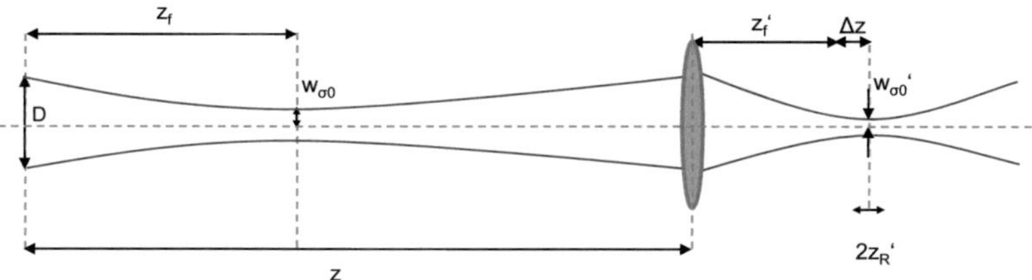

Figure 4.13: Parameters of an optical path. Here the beam is supposed to be already focused when leaving the laser resonator. The first beam waist $2w_{\sigma 0}$ is located at distance z_f. The effectice focusing optic is located at distance z - the length of the optical path. The focus distance of this optic is shifted and the focus lies at $z'_f + \Delta z$.

This theoretical spot size of the optical path $w'_{\sigma 0}$ could be verified experimentally: ablation craters in the focal distance show a diameter of $200\,\mu$m. Beyond the beam waist the crater diameter is supposed to increase. Since the intensity is distributed on a larger area, the observable crater diameter is decreasing after exceeding a particular distance from the beam waist due to the energy, which is not sufficient anymore to promote bone ablation all over the irradiated area.

4.4 Influencing Parameters

Multiple parameters effect the laser bone ablation process. With regard to the computer and robot assistance and the according preoperative planning, the inclination angle, focal distance, depth, and fluids which catalyze the process are of most importance. Therefore these parameters were analyzed[1] in the scope of this doctoral thesis and described in the following.

---

[1]These results are partly published in [Bur09a].

4.4.1 Angle

f-θ lenses show a planar focal field but varying incidence angles, which are dependent on the scanning angle of the scan head. Within limits of the small angle approximation the overall scanning angle is given by

$$\theta \approx \sqrt{\theta_x^2 + \theta_y^2},$$

while the scanning angle in x- and y-direction is given by

$$\theta \approx \tan\theta = P_{xy}/f$$

with P_{xy} is the coordinate in the focal plane and f is the focal distance. Especially in the border areas of the scan field the focused laser beam is projected elliptically onto the focal plane. Figure 4.14 illustrate the effect of the incidence angle on the spot geometry. The laser beam has a normal incidence angle only if $\theta_x = \theta_y = 0$, i.e. in the center of the scan field. In any other location the incidence angle is unequal $0°$ resulting in an elliptical beam spot.

The spot area of the laser beam impinging a surface orthogonal is given by

$$A_{\sigma 0} = \pi \cdot w_{\sigma 0}^2.$$

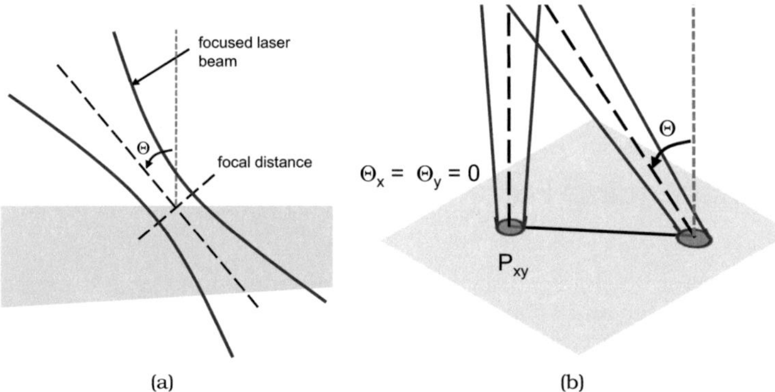

Figure 4.14: If a laser beam impinges the surface of a material orthogonal, the focus spot is circular. With an incidence angle θ the beam's focal plane is projected onto the surface and causes an elliptic focus spot.

If the laser radiation strikes the surface with an incidence angle θ the beam's focal plane is projected onto the surface. Thereby the area of the focal spot increases and the laser intensity decreases respectively. The area of the elliptical spot is given by

$$A_{xy} = \pi \cdot \frac{w_{\sigma 0}^2}{\cos\theta_x \cdot \cos\theta_y}.$$

The spatial distribution of the intensity in the scan field can be described as follows:

$$I(\theta_x, \theta_y) = I_0 \cdot \cos\theta_x \cdot \cos\theta_y,$$

with maximum I_0 is the intensity for $\theta_x = \theta_y = 0$. Figure 4.15 illustrates the spatial relation of the laser spot area and the laser intensity for the scan field of a two-dimensional scan head utilizing an f-θ lens.

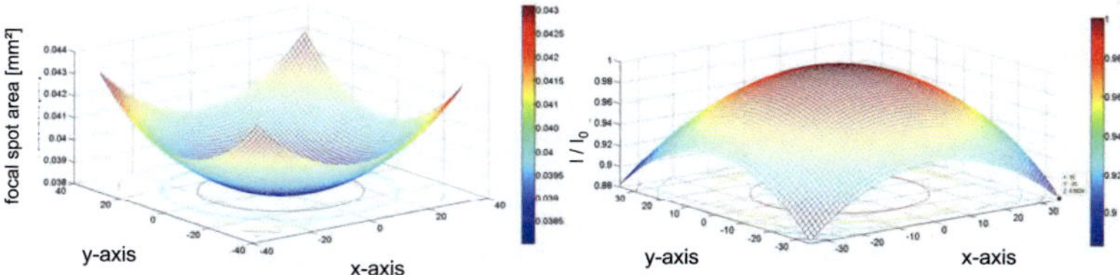

Figure 4.15: Area of the focal spot depending on the position in the scan field (left). Laser intensity distribution over the scan field (right). x- and y-axis values in mm. Colorbar values in µm.

In order to validate the hypothesis that the ablation geometry for laser pulses applied at the borders of the scan field changes according to the decreased intensity, an experiment was performed in the scope of a supervised student work [Mil09]. Ex-vivo marrow femur bone of a cow was cut in slices using a band saw and located in the focal distance under a two-dimensional scan head. Laser pulses were distributed over the scan field with constant energy per pulse. The resulting ablation craters were measured using the confocal microscope. The results are illustrated in Figure 4.16. The crater area is obviously directly correlated with the geometry of the laser spot. However, the effect of elliptical crater geometries becomes only significant in the border of the scan field. Therefore the scanning angle may be restricted in order to neglect the elliptical spot projection and reduced ablation depth.

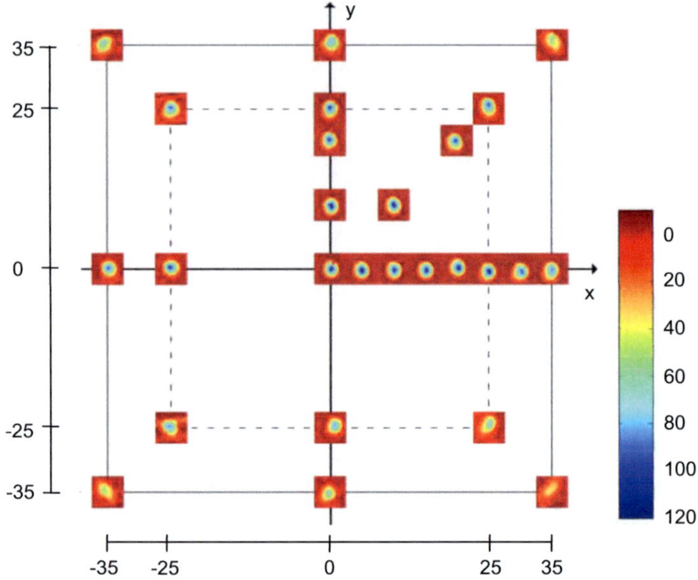

Figure 4.16: Top view on ablation craters distributed over the complete scan field of a two-dimensional scan head. In the center the crater is circular, while it becomes more and more elliptical at the borders of the scan field. The depths of the craters are scaled in μm and additionally color illustrated. Values of the x- and y-axis are stated in mm.

4.4.2 Distance

Corresponding to Section 4.3.4 and Equation 4.1 the tolerance volume of the scan head is given by the Rayleigh length. In order to validate this hypothesis for the removal of bone an experiment with bony specimen was performed. Fresh ex-vivo marrow bone of a cow's femur was cut in slices and adjusted in parallel to the scan head on a hexapod platform in the focal distance. Single laser pulses were applied in varying distances of the specimen from the focal distance ($\approx$ -2.5 mm to +10 mm). The experiment was repeated three times. Using the confocal microscope the resulting craters were measured and the depth was

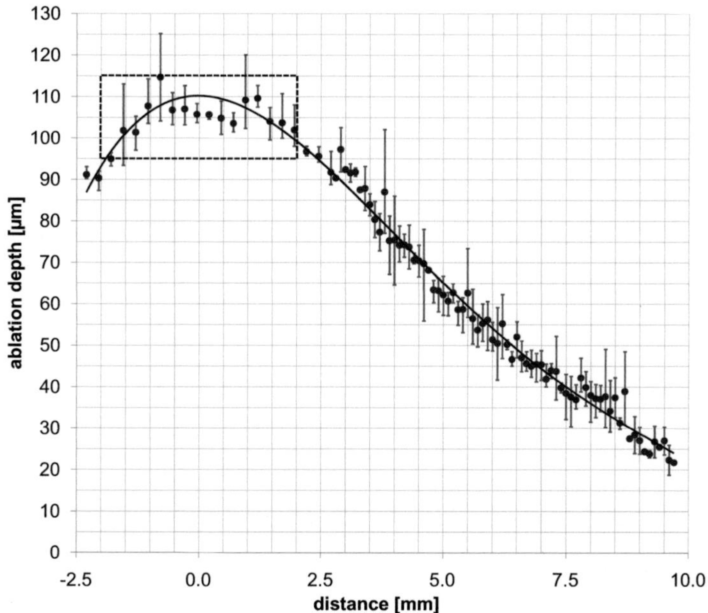

Figure 4.17: Mean depth of single laser pulse induced craters plotted against the distance of the specimen to the focal distance with fourth degree polynomial fitting function. Corresponding error bars show the maximum deviations of the mean value.

determined. The mean values for the depths are plotted against the distance in Figure 4.17.

The results proof that the maximal ablation depth is achieved, if the beam impinges the bone in its waist, i.e. the focal distance. Around the focal distance a tolerance area of $\pm$ 2 mm is defined. In this area the variance of the ablation depth is negligible ($\pm$ 10 μm). The tolerance area is slightly smaller than the Rayleigh length of 2.4 mm. Furthermore the experiment revealed that the ablation performance is nearly linear decreasing with increasing distance from the focal point. To conclude, the location of the focal point relative to the bone is a significant process parameter.

4.4.3 Depth

Cutting bony tissue by means of laser ablation necessitates knowledge of the depth development. The depth is corresponding to the number of pulses applied onto the same position. The functional dependency is determined experimentally. Fresh ex-vivo femoral cow bone was cut into parts and frozen. An hour before the experiment the bone was defrosted in water. The bone was fixated in modeling clay onto supporting material and adjusted by a hexapod in parallel to the scan head in the focal distance. Single laser pulses were applied with a pulse duration of 80 µs. The amount of laser pulses applied onto the same position was increased from one pulse up to one hundred pulses. The experiment was repeated 10 to 19 times for each amount of pulses.

Since ablation craters with a depth exceeding approximatly 1 mm could not be measured with the confocal microscope due to limited measurement range, the experiment was split up. Up to 10 single laser pulses applied onto the same position could be measured confocally with confidence. Figure 4.18(a) shows the single laser pulses applied onto bony specimen. For experiments with more than 20 laser pulses the bony specimen was precut using a band saw and single laser pulses were distributed along a cutting line over the gap (see Figure 4.18(b) for illustration) in order to determine the depth through the evaluation of the cross section. Precutting was necessary since slicing the bone after applying cutting lines would smear over the ablation with sawdust and distort the measurement. The depth of these line cuts was measured under a digital microscope.

(a)   (b)

Figure 4.18: Bony specimen for determining the depth development in dependency of the applied number of laser pulses. (a) Ablation craters produced by single laser pulses. (b) Precutted bone for evaluation under a digital microscope.

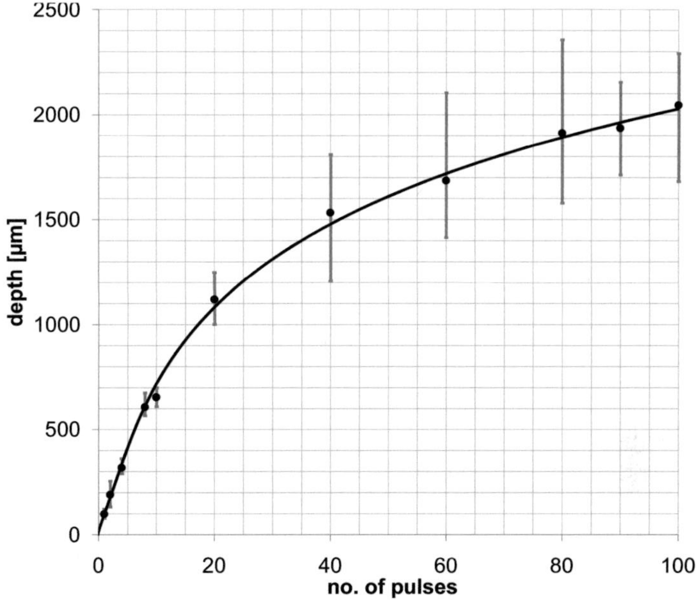

Figure 4.19: Mean resulting depth plotted against the number of single laser pulses (80 µs) applied with fitting function. Corresponding error bars indicate the maximum deviations of the mean value from the measurements.

The mean depth of the resulting ablation craters is plotted against the number of applied laser pulses in Figure 4.19. The maximum deviations from the mean value were determined. The larger deviations with increasing pulse amount is attributed to the less accurate measurement with the digital microscope. However, the analysis of the measurements proofed that the depth development is approximately logarithmic. The first laser pulses result in a fast depth development, since the laser energy is almost not attenuated. With increasing depth more and more energy of the laser beam is lost. On the one hand by the ablation debris and on the other hand by the increasing surface area of the incision [Iva00b, Iva05b].

4.4.4 Catalyzing Fluids

In the state of the art of short-pulsed CO_2 laser bone ablation, it was proven that applying a fine water spray onto the bone during laser ablation has a positive impact on the crater geometry and ablation depth (cp. Section 2.7.1). In the scope of this doctoral thesis almost every experiment was performed utilizing a fine distilled water spray continuously applied onto the bone.

However, in order to increase the ablation performance some experiments were conducted using alternative fluids replacing the water spray: pure glycerin, a glycerin water mixture, citric acid and a base. Because of the higher boiling point of glycerin compared to water, deeper ablation craters were expected through the resulting higher temperature. Experimental results revealed that glycerin or a glycerin water mix do not have any effect on the ablation depth. Additionally applying the acid or base on to the ablation site does not show any benefit compared to water.

In a further experimental series the chemicals were applied onto the bony specimen and preheated with a low energy laser pulse prior to the ablation pulse, which showed no effect as well. However, applying the chemicals and allow a longer soaking time induced nearly a doubling of the ablation depth. Furthermore with this longer soaking time, carbonization at the crater margins does not occur as it may be observed using the conventional water spray. A comprehensive summary of the experiments can be found in a joined publication with Mehrwald et al. [Meh10].

4.4.5 Thermal Dispersion

Ablation of vital hard-tissue necessitates to keep an eye on the thermal effects in order to prevent thermal side effects other than the intended thermo-mechanical ablation. Section 2.5.2 introduced the thermal interaction process taking place when irradiating tissue with laser. The thermal effects of laser bone ablation were considered in Section 2.7.2 and it could be shown that with specific chosen laser parameters no histological alarming impacts were observed.

However, regarding the process speed enhancement, increasing the repetition rate would allow time reduction. In this context knowledge about the exact temperature development inside the tissue becomes relevant, since it is the limiting factor. Therefore the thermal dispersion was analyzed in particular. Existing ablation models were reviewed (cp. Section 2.6.6). Since none of them is sufficiently describing hard tissue ablation using a short-pulsed laser, a comprehensive model was established: *Damping Ablation Velocity Model*. This allows simulation of the thermal dispersion into the bony tissue under irradiation with laser pulses. Figure 4.20 illustrates the temperature distribution after one laser pulse of 80 µs duration.

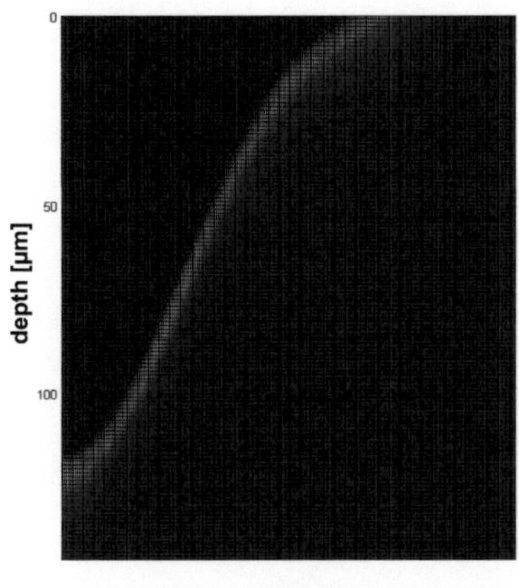

Figure 4.20: Simulated temperature distribution after application of a 80 µs pulse. The color bar indicates the temperature in °C. The value 0 reflects removed bone.

Based on the model it was possible to theoretically derive a safety distance between two succeeding laser pulses. However, these theoretical considerations necessarily have to be proven in histological tests with in-vivo specimen. The temperature as one of the main influencing process parameter is not considered for optimization further on in this doctoral thesis, but the topic is picked up in Chapter 8. The comprehensive consideration can be found in a related diploma thesis [Zha09].

4.5 Cutting by Pulse Concatenation

After investigation of the interrelationships for single laser pulses and their concatenation the composition of cuts is regarded in this section. The state of the art did not reveal the analysis of cuts based on the distribution of the single laser pulses. In fact, cuts were analyzed phenomenologically, i.e. for a specific set of laser parameters a cut is examined. Inspired by the single laser pulse analysis, the pulse concatenation for cutting techniques is investigated.

4.5.1 Theoretical Considerations

Bone ablation utilizing short laser pulses is by definition a discrete process. From the geometrical point of view the cut is composed of single discrete volume removals. The volume removal per pulse is corresponding to a 2d Gaussian function and, as described in the previous sections, dependent on the pulse energy and depth development. Hence, the cut can be described as the concatenation of single laser pulses. Figure 4.21 illustrates the theoretical ablation process for cutting. Repassing the laser beam in the second layer results in hitting the crater's surface of the previous pass. Hence, the resulting incision geometry is hilly.

In order to perform a cut, the scan head moves the laser focus with a constant velocity v_l along a predescribed trajectory. The pulse repetition rate R_f then affects the distance Δ between two consecutive pulses. Beside direct definition of v_l it is also possible, to define each laser pulse position in advance, specify the breaking distance between to succeeding laser pulses and thereby define v_l indirectly.

Since the volume removed by a single laser pulse corresponds to a 2d Gaussian profile, the cutting profile is by definition also of Gaussian shape (also termed wedge). However, since single laser pulses are distributed for shaping a cut, it will not be possible to achieve a straight cutting profile offhand as with a conventional cutting tool (i.e. saw).

4.5.2 Strategies

From the theoretical point of view there exist several strategies to distribute single laser pulses in order to achieve a specific cutting geometry. In the following the two most prominent techniques are presented.

Line Cut

In its standard definition a line cut is performed by the scan head with following a pre-described trajectory with a constant velocity. Since the laser is utilized short-pulsed with a specific repetition rate, single laser pulses are uniformly distributed along the trajectory according to that parameter. By repassing the trajectory, the cut is ablated layer by layer. The parameters of a line cut are the scanning velocity and the pulse repetition rate. Figure 4.22 illustrates schematically the line cutting technique.

In previous publications concerning pulsed laser bone ablation usually one line is executed. In the scope of this thesis widening of the cut is realized by definition of two (or more) adjoined line cuts according to the given trajectory (cp. Figure 4.22c). Furthermore the line is composed of single laser shots in the scope of this thesis. However, by defining single laser pulse positions along a predescribed trajectory leads to the same cutting technique as a conventional defined line cut.

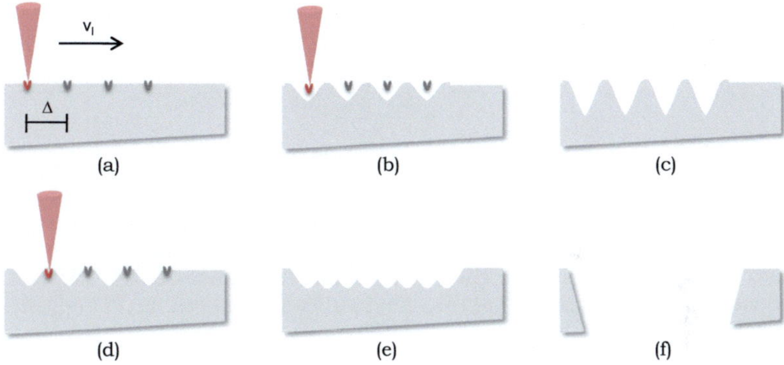

Figure 4.21: Single laser pulse concatenation. (a) The laser beam moves along a predescribed path with velocity v_l. (b) Repassing of the laser in the second layer causes the beam to hit the surface of the craters resulting from the previous pass. (c) After the second laser pass the volume removed corresponds to Gaussian profiles for each spot. (d) Lateral shifting of the laser pulse position before repassing lead to smoother incision geometries (e). (f) The cutting profile is typically shaped like a wedge.

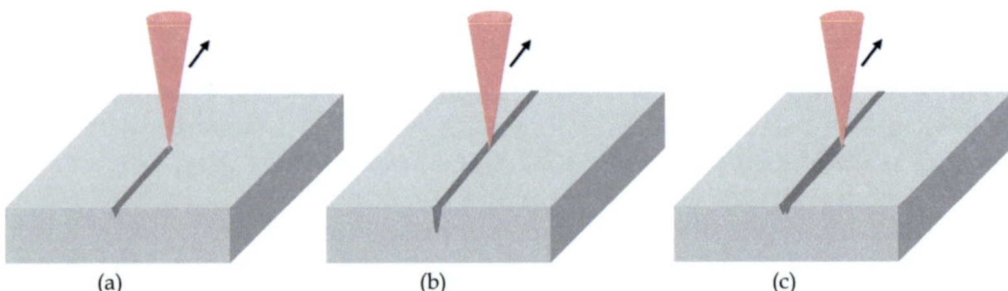

Figure 4.22: Line cutting strategy. (a) By moving the laser beam along a predescribed trajectory with a constant velocity the concatenation of single laser pulses results in a line. (b) Repassing the laser along the trajectory removes the bone layer by layer. (c) Widening of the cut by adjoining line cuts along the pre-described trajectory.

This approach is more reliable, since the synchronization of the laser trigger with the mirror movement cannot guarantee the absolute accurate position of each single shot when processing a line cut.

Another parameter specifying a line cut is the pulse overlapping factor n [Wer06], which is defined as follows:

$$n = \frac{R_f \cdot w'_{\sigma 0}}{v_l}.$$

The distance between two consecutive laser shots directly follows as $\Delta = v_l/R_f$ (cp. Figure 4.21a). This overlapping factor is usually considered when defining a line cut. However, the overlapping of pulses is not considered for each single line in the scope of this thesis, but between two consecutive passes of the laser beam. That means, that for each repassing the laser shots are relocated alternately about $\pm\Delta/2$ laterally.

Wobble Cut

The cutting depth for laser ablation is limited. This is due to the fact that the laser beam is more absorbed at the cut surface with increasing depth and additionally more attenuated by the debris within the incision. To overcome this drawback the cut necessarily has to be widened in order to reduce the effects causing depth limitation. Hence, the so called wobble technique is utilized [Iva05b]. Here the linear movement of the laser beam by the scan head is superimposed to a rotational movement (wobble) to widen the line cut. Figure 4.23 illustrates this technique schematically.

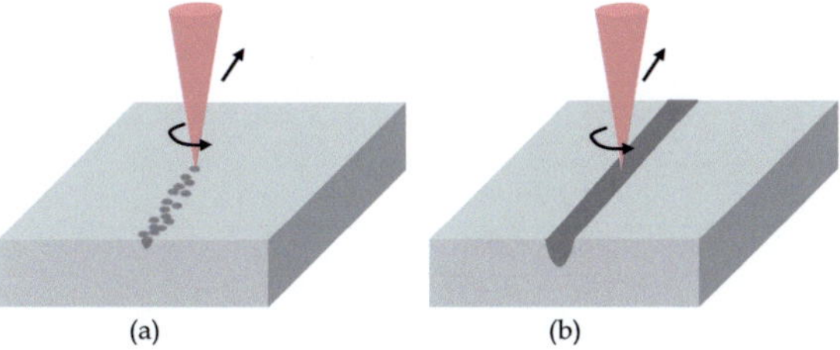

Figure 4.23: Wobble strategy. (a) The scanning is performed along a predescribed trajectory with a constant velocity superimposed with a rotational movement. Laser pulses are distributed according to the repetition rate. (b) The wobble strategy leads to a wider cut.

Wobble cuts are characterized as a line cut by the scanning velocity and the pulse repetition rate. The superimposed wobble movement is defined by the radius r_{wobble} and the frequency f_{wobble}. Figure 4.24 illustrates the distribution of laser pulses according to the wobble technique for two different parameter sets.

4.5.3 Experimental Evaluation

In order to validate the theoretical development of cuts according to the distribution of single laser pulses an experimental evaluation was performed. Fresh ex-vivo femoral cow bone was cut into parts and frozen. An hour before the experiment the bone was defrosted in water. The bone was fixated in modeling clay onto supporting material and adjusted by a hexapod in parallel to the scan head in the focal distance. Further descriptions of these experiments could be found in an associated diploma thesis [Pla09].

Line Cut

Figure 4.25 shows cutting profiles of line cuts performed with 80 µs pulse duration. With increasing cutting depth the cut widens noticeably. Furthermore the cutting profile becomes more narrow and shows the typical wedge geometry.

Publications describing the state of the art of hard tissue ablation usually show cutting results on the basis of cutting profiles. For the first time the cutting profile was analyzed in cutting direction in the

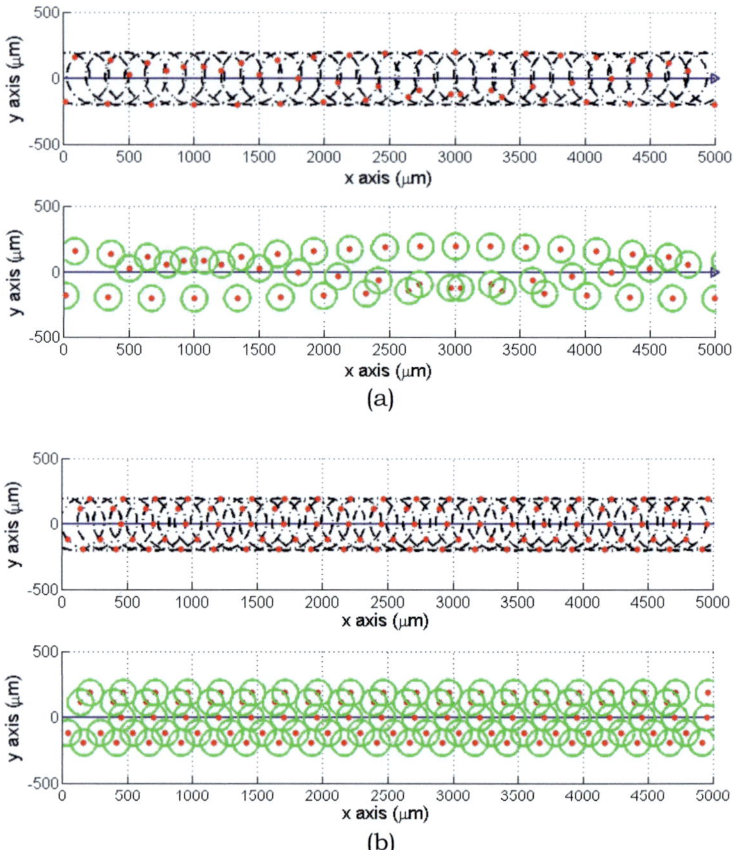

Figure 4.24: Wobble technique to superimpose linear movement (r_{wobble} = 200, v_l = 20 mm/sec) with different parameters. (a) Scanning movement (top) for f_{wobble} = 135 Hz along a pre-described trajectory (purple) and laser pulse distribution for R_f = 200 Hz (red dots). Corresponding cut geometry (bottom) with ablation radius $w'_{\sigma 0}$ = 100 µm. (b) f_{wobble} = 160 Hz, R_f = 400 Hz.

scope of this doctoral thesis using a confocal microscope. Therefore a line cut with an overall length of 600 µm was defined. The first layer consists of two pulses at location 200 µm and 400 µm. In the consecutive layer the pulses are relocated about $\Delta/2 = 100\mu m$ leading to a layer of three pulses at 100 µm, 300 µm and 500 µm. This is corresponding to a pulse overlapping factor of 1. Up to twelve consecutive lines were processed. Figure 4.26 shows the confocal measurement for three, six and twelve lines.

After three consecutive lines the profile along the cutting directory shows three peaks corresponding to the Gaussian profiles of the single pulses. The length of the cut is 600 µm at the surface and 400 µm at the bottom. After six consecutive lines the bottom of the cut becomes more smooth thanks to the overlapping. However, the length of the cut measured at the bottom decreases noticeably while the edges become more and more slanted. This becomes particularly noticeable

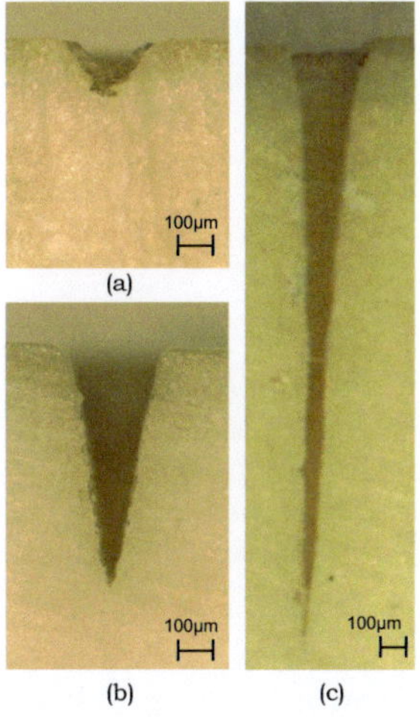

Figure 4.25: Microscopic images of line cuts with focal position at the surface at 80 µs pulse duration. (a) After one pulse the ablation depth is 117.84 µm and a width of 244.05 µm. (b) Ten consecutive pulses result in an ablation depth of 673.54 µm and a width of 265.92 µm. (c) After one hundred consecutive pulses an ablation depth of 2158.74 µm is reached. The cutting width increased to 286.57 µm.

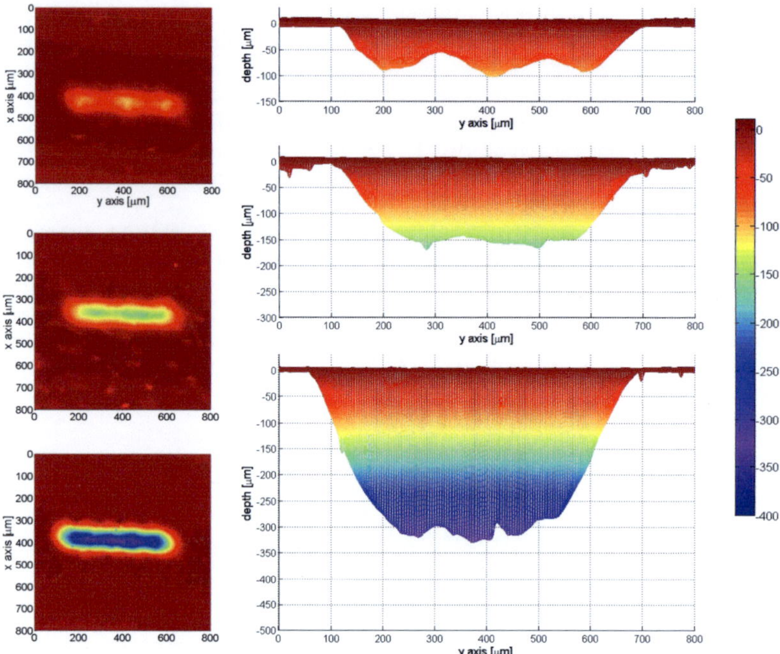

Figure 4.26: Confocal measurement of line cuts with 80 µs pulse duration and alternately relocated pulses about $\pm\Delta/2 = 100\mu m$ (right: top view, left: profile in cutting direction). Top: Three consecutive lines; Middle: Six consecutive lines; Bottom: Twelve consecutive lines.

after twelve consecutive layers. As stated in Section 4.5.1 a laser cut cannot be processed with straight edges. Even applying additional laser pulses to the emerging edges will always lead to slanted edges.

Wobble Cut

A wobble cut was performed with 80 µs pulse duration and 80 consecutive lines. Figure 4.27 shows the cutting profile for a wobble cut with $r_{wobble} = 200\,\mu m$, $v_l = 40\,mm/sec$, $f_{laser} = 200\,Hz$ and $f_{wobble} = 135\,Hz$. The cutting depth is 1896.73 µm and width 609.54 µm. In comparison with line cuts the cut is wider and shows a less narrow profile in depth.

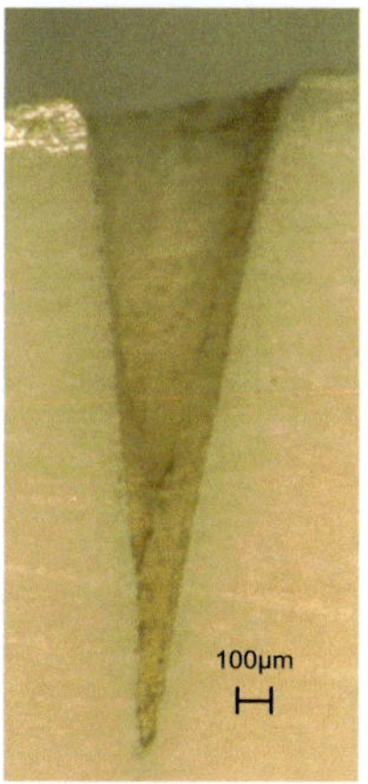

Figure 4.27: Microscopic image of an exemplary wobble cut profile after 80 consecutive lines with focal position at the surface at 40 µs pulse duration, r_{wobble} = 200 µm, v_l = 40 mm/sec, f_{laser} = 200 Hz and f_{wobble} = 135 Hz. The cutting depth is 1896.73 µm and width 609.54 µm.

Furthermore a wobble cut with an overall length of 600 µm was defined. Three consecutive lines were processed with 80 µs pulse duration, r_{wobble} = 200 µm, v_l = 20 mm/sec, f_{laser} = 200 Hz and f_{wobble} = 135 Hz. Figure 4.28 shows the confocal measurement for one, two and four layers. The faster depth development is due to the fact that one layer contains more laser pulses. The cut is wider according to the wobble radius. Additionally the bottom of the cut is sooner becoming smooth. However, the faster depth development is gained by a wider cut.

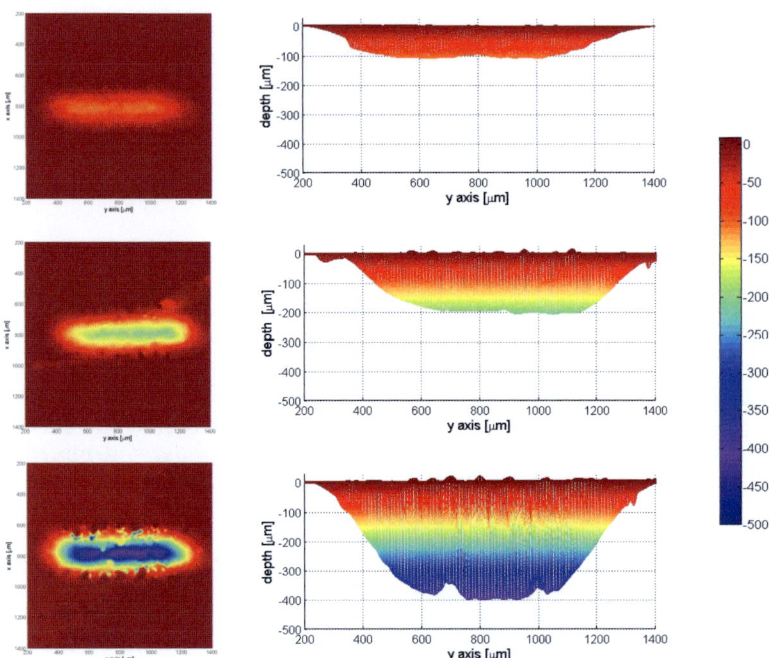

Figure 4.28: Confocal measurement of wobble cuts with $80\,\mu\text{s}$ pulse duration, $r_{wobble} = 200\,\mu\text{m}$, $v_l = 20\,\text{mm/sec}$, $f_{laser} = 200\,\text{Hz}$ and $f_{wobble} = 135\,\text{Hz}$. (right: top view, left: profile in cutting direction). Top: One layer; Middle: Two consecutive layers; Bottom: Four consecutive layers.

4.6 Conclusion

Laser bone cutting as a discrete process necessitates by definition a process model based on single laser pulses and their concatenation. For the first time a systematic modeling of the short-pulsed laser ablation process with a two-dimensional optical scan head was performed in the scope of this doctoral thesis.

It could be shown, that by restricting the scanning angle the inclination angle can be neglected in a reduced scan field. The hypothesis that the tolerance area around the focus, in which the variance in the resulting ablation crater is negligible, corresponding to the Rayleigh distance of the utilized laser, could be verified experimentally and thereby confirmed for the first time.

The depth development for the concatenation of single laser pulses was evaluated experimentally and a logarithmic related behavior was shown. For the first time the variance in the ablation depth was evaluated on the basis of 211 measured single laser pulses. The results revealed that the variance could be assumed as Gaussian distributed and that the ablation depth varies between -3% and +7% from the mean ablation depth. However, it will be necessary to correlate these results not only to the number of applied pulses but also to the incision depths, widths and repetition rate in further investigations.

Cutting techniques were considered theoretically on the basis of the distribution of single laser pulses and experimentally evaluated. For the first time cuts performed with laser ablation were evaluated in their profile in cutting direction. Furthermore it was revealed that straight cutting profiles cannot be obtained offhand with laser ablation. In the scope of this thesis, cuttings are performed by placing successive laser pulses next to each other, which is similar to the described line cutting technique. The pulse repetition rates of up to 400 Hz are chosen out of the literature in order to achieve histological compatibility. However, the results of Zhang indicate that the repetition rate can be quintupled when adapting the distance between successive laser pulses according to thermal considerations [Zha09]. This hypothesis has to be proven in in-vivo experimental series and histological tests.

The findings and parameters described in this chapter are the basis for computer assisted planning for robot assisted laser osteotomy. The understanding of short-pulsed laser ablation for cutting bony tissue as a discrete process allows accurate planning of bone removal based on the concatenation of single laser pulses and prediction of the cutting depth in advance.

5

Planning Robot Assisted Laser Ablation

The task of determining a sequence of actions that will achieve a specified goal is called planning. Generally a planning problem inherits a state space, which is often too large to be represented explicitly and is therefore often implicitly described by the planning algorithm. If the state space is finite or computably infinite the planning problem is discrete.

In order to plan, an initial state and a goal state need to be defined. A finite number of actions are then applied to manipulate the state. The concatenation of actions results in a sequence of states, that represents the plan. The plan can either follow a strategy or be reactive. There are two criteria for planning: feasibility and optimality. While feasible planning strategies determine a plan regardless of its efficiency, optimal planning strategies find a feasible plan that optimizes a performance index [LaV06].

In case of laser ablation we are facing a discrete process, i.e. single laser pulses which have to be applied onto bone so that a cut is achieved. In contrast to industrial laser cutting processes the target in the scope of this doctoral thesis is a human. Bone is an individual workpiece. Hence, planning aims at achieving a feasible laser cut for a specific patient. First, the cut (osteotomy) has to be defined geometrically. Then this geometric definition has to be transferred into an ablation pattern, i.e. a distribution of single laser pulses. This ablation pattern is the basis for the following execution planning, which aims at

determining optimized locations for the scan head. The working volume
of the scan head and laser process parameters, such as optimal ablation near the beam waist and the inclination angle, have to be taken
into account, thus becoming a constrained optimization problem. In
the following the developed methods for geometrical planning and the
transfer into an ablation pattern are described. Furthermore the optimization algorithms for achieving scan head locations are explained.

5.1 Cutting Geometry Planning

The first step towards robot assisted laser ablation is the definition
of the ablation geometry. In the scope of this doctoral thesis cutting
paths and cutting trajectories were specifically addressed. Figure 5.1
illustrates the process of geometry planning. Cutting geometry planning
starts with the definition of the cutting path. By specifying the cutting
angle and cutting depth, a cutting trajectory is defined. In the following
sections cutting geometry planning is explained.

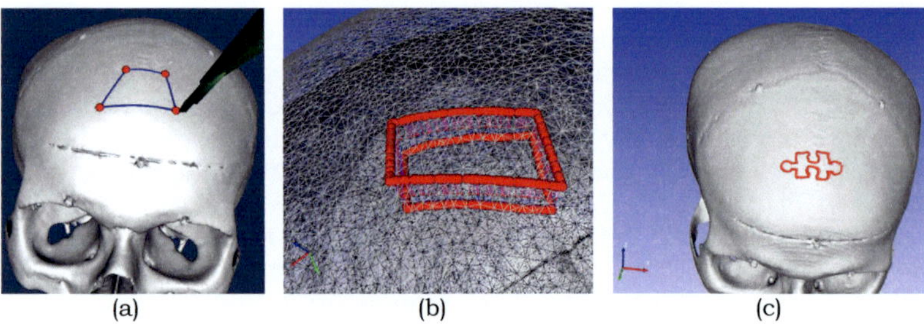

(a)   (b)   (c)

Figure 5.1: Geometry planning. (a) After defining support points on the
surface, waypoints are calculated according to an interpolation function (here piecewise linear interpolation). (b) The
cutting depth is determined by projection along the cutting normal and determination of corresponding points on
the underside of the bone. (c) Exemplary complex cutting
geometry.

5.1.1 Cutting Path

The geometrical definition of a cut starts with specifying the three-
dimensional path. Along this path the cut has to be processed. If no

cutting depth is specified, the cutting path corresponds to a marking. A cutting path is generated in two steps. First, the support points are defined and second, waypoints in between are generated according to an interpolation function.

Definition of Support Points

A set of support points $S = \{s_i, i = 0..n\}$ is defined on a three-dimensional triangulated surface model. Hence, a support point s_i is described by its position on the surface model $p_i = (x, y, z)$. A cutting path is called closed, if $s_0 = s_n$ and called open if $s_0 \neq s_n$.

Interpolation

Based on the set of support points S the cutting path is defined by choosing an interpolation type. According to the interpolation function a set of waypoints $W = \{w_j, j = 0..m, m \geq n\}$ is determined, with W is a superset of S. Hence, the following applies

$$S \subseteq W \Leftrightarrow \forall x \in S : x \in W.$$

The following interpolation types can be utilized:

- *Linear interpolation* on a set of support points is defined as the concatenation of linear interpolation functions between succeeding pairs of these points.

- *Polynomial interpolation* is a generalization of linear interpolation. Instead of using a linear function a polynomial of higher degree is used as interpolation function. Given n support points, there is one polynomial of degree at most n-1 interpolating all points.

- *Spline interpolation* avoids disadvantages of polynomial interpolation, i.e. computational expensiveness and exactness. Instead of searching for a high-degree polynomial interpolating the given support points, spline interpolation utilizes low-degree polynomials for piecewise interpolation and additionally assures smoothness between the pieces.

- *Variations:* Piecewise interpolation or combinations of interpolation functions can be utilized in order to define more complex cutting shapes.

The amount of waypoints depends on the resolution of the surface model. Assuming a trinagular mesh representation of the surface and depending on the interpolation function, a waypoint is set on each passed triangle edge. The higher the density of the mesh the more waypoints will be generated. Obviously, the overall accuracy of the complete laser osteotomy is strongly influenced by the quality of the surface model of the patient. This is in turn restricted by the image acquisition method, the segmentation and triangulation quality. Hence, the accuracy of the patient model is an important process parameter.

5.1.2 Cutting Trajectory

A cutting trajectory is based on a cutting path definition. Additionally, the definition comprises a specific cutting depth and a cutting angle. Usually the cutting depth corresponds to the bone thickness, i.e. the cut is supposed to go through the bone in order to perform an osteotomy.

Cutting Angle

For each waypoint the surface normal of the corresponding triangle cell is determined (cross product of two triangle edges) to indicate the cutting angle, which is at first supposed to be normal to the bone surface. Hence, a waypoint w_j is described by its position on the surface model $p_j = (p_x, p_y, p_z)$ and its normal $n_j = (n_x, n_y, n_z)$.

The cutting angle α specifies the angle between the surface normal $\vec{n}$ and the cutting direction $\vec{c}$. A cone is spanned by α around the surface normal. The cutting direction $\vec{c}$ is defined by the polar angle θ which indicates the radial coordinate on the circular base of the cone. θ is defined anticlockwise from the positive x-axis, i.e. the polar axis, corresponding to the global planning coordinate system. Figure 5.2 illustrates the definition of the cutting angle and cutting direction.

Cutting Depth

After the definition of the cutting angle or cutting direction respectively the cutting depth is the remaining parameter describing a cutting trajectory. The cutting depth is usually corresponding to the bone thickness, i.e. the cut is supposed to go through the bone in order to perform an osteotomy. Here, the cutting depth is determined by the projection along the cutting direction of each waypoint w_j and determination of the corresponding point c_j on the bottom of the bone by intersection with the surface mesh. If no separating cut is planned, the cutting depth is defined from the cutting path in cutting direction and corresponding lower waypoints are set. Figure 5.3 illustrates a cutting trajectory.

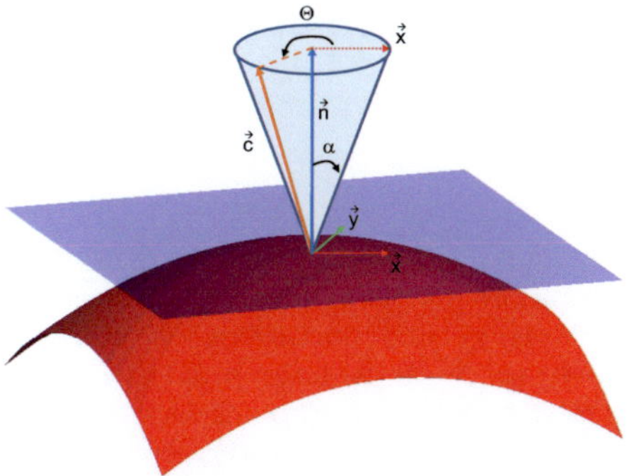

Figure 5.2: The cutting angle α spans a cone around the surface normal (blue). The polar angle θ defines the cutting direction (orange) anticlockwise from the positive x-axis (polar axis).

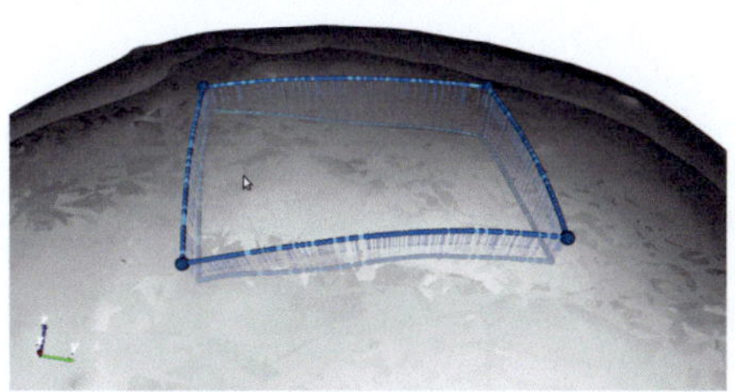

Figure 5.3: A cutting trajectory is defined by a set of support points (dark blue) and waypoints (light blue) according to an interpolation function. Each waypoint on the surface has got a corresponding point on the underside of the bone defined by the cutting direction (arrows) and thereby indicates the cutting depth at this specific point.

5.2 Transfer into Ablation Sequence

Short-pulsed CO_2 laser ablation is by definition a discrete process. As described in the previous chapter the concatenation of single laser pulses in a specific distribution generates a cut. Hence, the cutting geometry as described in Section 5.1 has to be transformed into a set of single laser pulses in order to be processed.

5.2.1 Constraints

The planning constraints are directly derived from the discrete process model for laser ablation (cp. Chapter 4). Depending on the chosen cutting strategy the geometrical cutting trajectory description is transferred into a discrete set of single laser pulses. The cutting depth and laser parameters have to be taken into account as constraints, in order to derive a feasible laser pulse set for the given geometry.

5.2.2 Transfer of a Cutting Geometry into an Ablation Pattern

According to the chosen cutting strategy and the geometrical definition of the cutting path, single laser pulses are distributed along the cutting path considering the cutting direction which is corresponding to the laser beam inclination angle. Hence, each laser pulse is represented by its location, i.e. position p_i and inclination direction c_i with $i \in q$ and q is the total amount of single pulses.

In a first step single laser pulses are distributed along the cutting path. Depending on the cutting strategy (cp. Section 4.5) the pulse overlapping factor or number of lines is given. Thereby the first layer of laser pulses is determined. In case of a marking path the ablation pattern is complete.

Corresponding to the findings of Section 4.4.3 the ablation depth is dependent on the number of pulses applied. The inverse function of the fitting function from Figure 4.19 is utilized for determining the amount of pulses q necessary in order to achieve a specific cutting depth x:

$$q(x) = a \cdot \exp(-b \cdot x) + c \cdot \exp(-d \cdot x) + e \cdot \exp(-f \cdot x) + g$$

with $a = 419.5$, $b = 0.00977\,\mu\mathrm{m}^{-1}$, $c = -421$, $d = 0.00974\,\mu\mathrm{m}^{-1}$, $e = 4$, $f = -0.0016\,\mu\mathrm{m}^{-1}$ and $g = -2.6$. Figure 5.4 illustrates $q(x)$.

This equation can be regarded as an approximation function considering concurrent processes during laser ablation, i.e. attenuation of

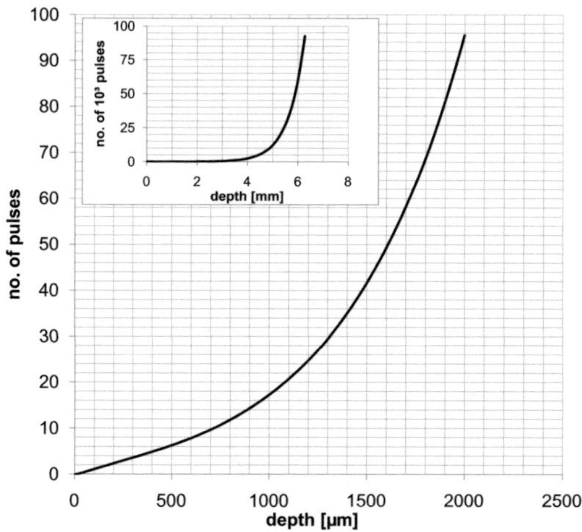

Figure 5.4: Number of 80 µs pulses plotted against the depth.

laser intensity by debris and water spray, loss of laser energy at the crater walls with growing depth, catalytic effect of water and laser spot expansion with growing distance from focus. Figure 5.4 indicates the asymptotic behavior of laser bone ablation at approximately 8 mm cutting depth. However, this approximation function was determined for a pulse duration of 80 µs. Shorter or longer pulse durations result in an additional factor respectively other coefficients in the approximation function.

Coming back to the transfer of a cutting trajectory into an ablation pattern. Now that the depth is represented by a function which calculates the corresponding number of laser pulses, a cutting trajectory can be filled up with laser pulse position according to a cutting strategy. The cutting depth is implicitly given by the distance between all pairs of waypoints (one at the surface of the incision and one at the bottom). Laser pulse positions are situated along the cutting path layer by layer. For each successive layer, further pulse positions are projected in cutting direction and according to the depth development function, untl the specified cutting depth is reached. Figure 5.5 exemplarily shows the transfer for a given cutting trajectory (cp. Figure 5.3) into an ablation pattern for a line cut with no overlapping.

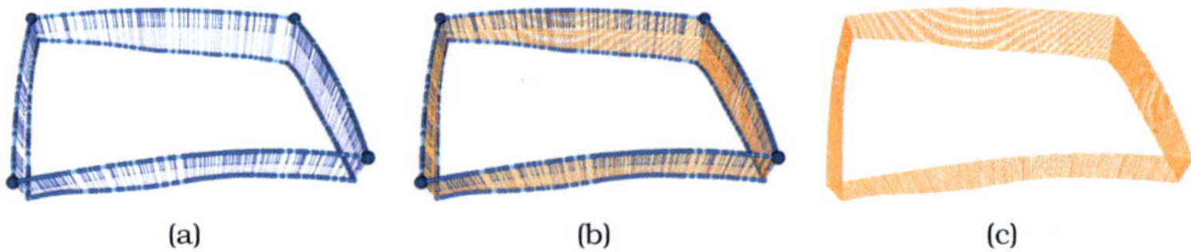

(a) (b) (c)

Figure 5.5: (a) The predefined cutting trajectory is composed of one cutting path on the surface and one at the underside of the bone. Each waypoint has a corresponding waypoint on the underside of the bone, hence, the cutting depth is given at this specific point. (b) According to the process model, the geometrical representation of the cutting trajectory is now filled with laser pulse positions according to the cutting strategy. (c) After the transfer into an ablation pattern, the geometrical information is implicitly represented by the laser pulse positions.

5.3 Optimal Execution Planning

In the previous section, the transfer of the cutting geometry into an ablation pattern, consisting of laser pulse positions, was performed without taking the way of pulse application into account. In fact, the transfer is based on the findings for depth development and cutting strategy. Hence, the relation of pulse positions to the actual application is considered in the scope of optimal execution planning. In the scope of this thesis two-dimensional optical scanning is considered (cp. Section 4.3.4). Thus, for processing the ablation pattern, corresponding locations of the scanning device (scan head) have to be determined which cover the cutting path.

5.3.1 Problem Formulation

The working volume of the scan head may not be sufficient for executing all laser pulses of an ablation pattern out of one location. This can be simply due to the fact that the cutting geometry exceeds the working volume on the one hand. Or on the other hand, due to the fact that inclination angle of the laser beam for a specific pulse position resulting from one scan head location may not be optimal in conjunction with the curvature of the bone. Hence, execution planning requires the determination of scan head locations for a given ablation pattern under constraints, such as working volume and inclination angle, thus becoming a constrained optimization problem. Towards robot assisted execution, the number of repositionings of the scan head should be reduced. Thus, the solution of the optimization problem has to ensure a minimal amount of scan head locations. This optimization problem is illustrated in Figure 5.6 in the two-dimensional case. The two-dimensional path has to be covered by the rectangular working area of the scan head. Obviously the solution illustrated in Figure 5.6(b) is not optimal regarding the number of rectangles. Figure 5.6(c) shows one optimized solutions where the total number of rectangles could be reduced to a minimum.

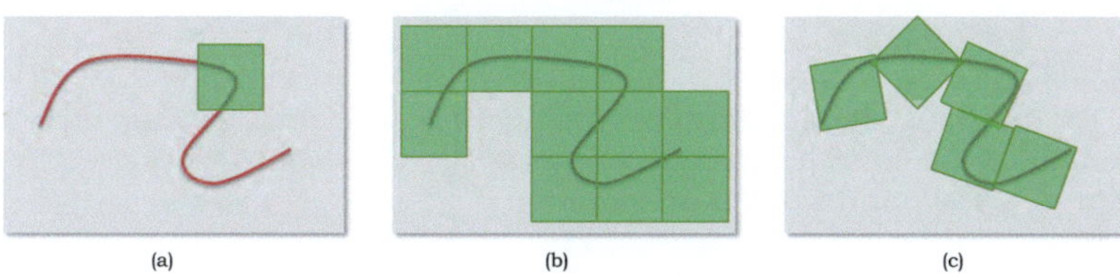

Figure 5.6: (a) The cutting path (red) and one scan head location with its corresponding rectangular working area (green) covering a part of the cut. (b) Partioning the solution space in equidistant sets and choosing all sets which contain parts of the cut does not result in an optimal solution. (c) Optimized solution for the cutting path contains the minimum number of scan head locations.

The *minimum set cover problem* is a prominent optimization and combinatorial problem, since it is NP-complete. Richard Karp mentioned it together with 20 other NP problems in his famous paper 1972 [Kar72]. Classically, the minimum set cover problem is defined as follows:

Instance A finite set S and collection C of subsets of S.

Solution A set cover of S, i.e. a subset $D \subseteq C$ such that each element of S is contained in some set in D.

Measure $|D|$.

The set-covering problem abstracts many commonly arising combinatorial problems [Cor01]. This decision problem generalizes the NP-complete vertex cover. Thus, it is also NP-hard. The problem of finding the minimum amount of scan head locations covering a given ablation pattern, can be formulated as a minimum set cover problem as follows:

Instance An ablation pattern P (either in 2d or 3d) can be represented by a finite number of points w_i with $i = 1..n$, each representing a laser pulse position. Let $F = \{L_1, L_2, ..., L_m\}$ be a family of subsets of P such that $\bigcup_{i=1}^{m} L_i = P$. Each subset L represents laser pulse positions covered by a specific scan head location.

Solution A collection of subsets $C \subseteq F$ is a set cover of P if $P = \bigcup_{S \in C} S$.

Measure The size $|C|$ of a cover C is the amount of subsets in C, i.e. the amount of scan head locations. Given P and F, the optimization algorithm solving this problem should output a set cover C with minimum size.

To conclude, the optimization problem requires an algorithm which computes a solution with measure as small as possible for a given instance of the minimization problem.

5.3.2 Two-dimensional Solution

For solving a minimum set cover problem approximation algorithms are the method of choice. Assuming that if the size of the instance increases, the size of the approximation grows relative to the size of an optimal solution, the approximation ratio is logarithmic. Thus, applying a greedy heuristic may give suboptimal but good results [Cor01]. Hence, in the scope of this thesis the problem is solved by a greedy algorithm.

The greedy method is rather intuitive. At each stage, the set $S \in F$ is added to the solution that covers the greatest number of remaining elements that are uncovered [Cor01]. Algorithm 1 implements the greedy minimum set cover solution.

The greedy algorithm is an asymptotically best possible approximation algorithm for the minimum set cover problem (see proof in [Cor01]). However, the found scan head locations need to be ordered for the robotic execution. The locations are therefore ordered (cp. Travelling Salesman Problem).

The quality of the solution is obviously dependent on the family of subsets F which span the solution space. Here, the solution space is given by all possible scan head locations $L_j = (x, y)$, which cover at least one laser pulse of the ablation pattern $P = \{w_1..w_n\}$. Hence, the solution space is given by the convex hull $H_{convex}(P)$ of P enlarged by half of the scan field length. The family of subsets F is derived from the solution space by equidistant spacing.

Require: Finite set P and family of subsets F
Ensure: Minimum set cover C
  $C \leftarrow \emptyset$
  $U \leftarrow P$ {U is the set of all uncovered elements}
  while $U \neq \emptyset$ **do**
    Choose $S \in F$ with $\max |S \cap U|$ {S covers largest number of elements in U}
    $C \leftarrow C \cap \{S\}$ {add S to the cover}
    $U \leftarrow U \backslash S$ {remove all elements from U covered by S}
  end while
  return C

Algorithm 1: Greedy algorithm solving the minimum set cover problem.

Suppose that the finite set P is ordered, i.e. the first element in P is the first laser pulse to be shot. A modified greedy algorithm (see Algorithm 2) solving the minimum set cover problem does not start with an explicitly given solution space F. In fact, subsets of F are spanned in each iteration step. The modified greedy algorithm chooses the first element out of P and spans the solution space around this element by creation of an equidistant grid with dimension dim. All elements S_i out of the solution space F are candidates, which are evaluated according to the amount of elements out of U covered by S_i. The algorithm chooses S with the largest number of elements covered and adds S to

---
Require: Finite set P (ordered)
Ensure: Minimum set cover C
  $C \leftarrow \emptyset$
  $U \leftarrow P$ {U is the set of all uncovered elements}
  while $U \neq \emptyset$ **do**
    $loc = p_0 \in U$ {select first element of uncovered elements}
    $F = GRID_{equidistant}(loc, dim)$ {span solution space around loc}
    Choose $S \in F$ with $\max |S \cap U|$ {S covers largest number of elements in U}
    $C \leftarrow C \cap \{S\}$ {add S to the cover}
    $U \leftarrow U \backslash S$ {remove all elements from U covered by S}
  end while
  return C

Algorithm 2: Modified greedy algorithm for solving the minimum set cover problem.

the minimum set cover C. All covered elements are removed from U and the next iteration step.

The modified greedy algorithm assures that the scan head locations are situated along the cutting path direction. Therefore the solution is already providing locations which are optimal for the robotic execution.

5.3.3 Three-dimensional Solution

The minimum set cover solution can be transferred into the three-dimensional case. The working area of the scan head is now a cube. In order to account for the rotation symmetry of the scan head, the working area may also be regarded as a frustum. An illustration of the three-dimensional problem is depicted in Figure 5.7.

Figure 5.7: Illustration of the three-dimensional minimum scan head location problem for a given cutting trajectory.

Kinematic Modeling of the Scan Head

While the two-dimensional case with a quadratic scan field allows easy decision, whether a laser pulse is within or not within the working field, the three-dimensional case necessitates modeling of the working area of the scan head. As described in Section 4.3.4 the working area is corresponding to the frustum of a pyramid. In the scope of this doctoral thesis the scan head is modeled kinematically:

- two perpendicular revolute joints $\theta_{x/y}$ for considering the scanning angle

- one revolute joint φ to account for the symmetry of the scan area around the laser beam

- one prismatic joint to model the tolerance area

Figure 5.8a illustrates the working volume of the scan head, which is corresponding to the frustum of a pyramid. In order to simplify the planning of scan head locations the rotation symmetry of the scan area is considered. That leads to a frustum workspace (cp. Figure 5.8b). The kinematic chain is illustrated in Figure 5.9. The Denavit Hartenberg parameters for describing the kinematics are given in Table 5.1. $\theta_{x/y} \in \{0..max_{angle}^\circ\}$ indicate the scanning angle, $\varphi \in \{0..360^\circ\}$ reflects the rotation symmetry and $d_t \in \{0..2t\}$ (t = tolerance area). Determination of max_{angle} and t is described in Section 4.4.1 and 4.4.2.

Table 5.1: DH parameters for a two-dimensional scan head. With $\theta_{x/y} \in \{0..max_{angle}^\circ\}$, $\varphi \in \{0..360^\circ\}$ and $d_t \in \{0..2t\}$ (t = tolerance area)

i	θ_i	α_{i-1}	a_{i-1}	d_i
1	θ_x	0	0	0
2	θ_y	-90	0	0
3	φ	90	0	0
4	0	0	f	d_t

Determination of Optimal Scan Head Locations

The two-dimensional minimum set cover problem can be transferred into the three-dimensional case. The cut is represented by the ablation pattern defined by n locations L, each given by its position $p_L = (x, y, z)$ and the direction $n_L = (a, b, c)$. Each of these locations represents a single laser pulse. The original and modified greedy algorithm (cp. Algorithm 1 and Algorithm 2) can be adapted for finding an asymptotically best possible approximation algorithm for the minimum set cover problem.

Taking into account the characteristics of the working area of the scan head (cp. Figure 5.8), the rotation symmetry of the laser beam constraints the orientation of the scan head in two degrees of freedom. One degree of freedom remains, i.e. the orientation around the laser beam axis. Furthermore, the kinematic modeling of the two-dimensional scan head facilitates the decision whether a laser pulse is executable in the current scan head location or not.

Here, the solution space is given by all possible scan head locations $L_j = (x, y, z, a, b, c)$, which cover at least one laser pulse w_i out of the ablation pattern $P = \{w_1..w_n\}$. Hence, the solution space is given by the oriented bounding box of P enlarged by half of the scan field length in x- and y-direction and the tolerance area in z-direction of the coordinate

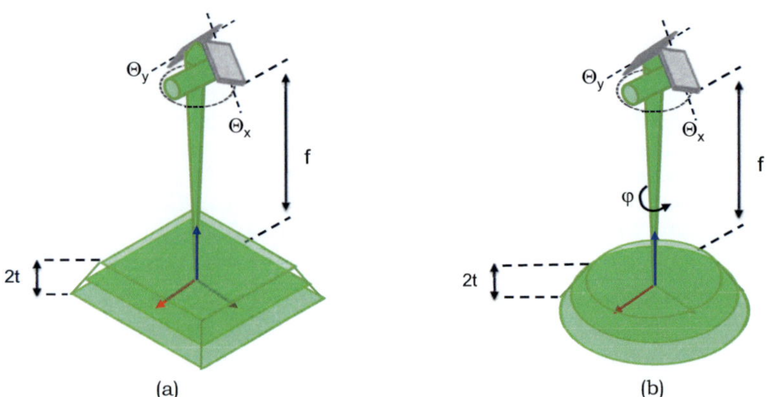

Figure 5.8: (a) The scan field is generally quadratic. Considering the tolerance area around the focal plane in which optimal ablation takes place, leads to a workspace corresponding to the frustum of a pyramid. (b) Considering the rotation symmetry of the laser beam leads to a frustum workspace.

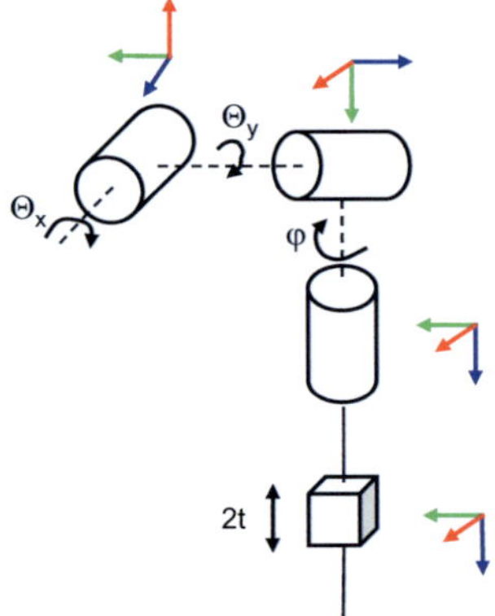

Figure 5.9: Kinematic chain for the scan head. Two revolute joints for the mirrors, one revolute joint to account for the rotation symmetry of the beam and one translational joint for the tolerance around the focal point.

system of the bounding box. The family of subsets F is derived from the solution space by equidistant spacing. The dimension of this spacing should be smaller in z-direction, i.e. the position of the focal plane which accounts for the focal tolerance. Figure 5.10 illustrates the solution space determination. The solution space is visualized with colored spheres. Each sphere indicates a possible position of the scan field center for the corresponding scan head location. The color gradient from red to green and growing size of the spheres reflect better rating, i.e. the percentage covered from the cutting trajectory in the specific scan head location.

Whether a covered single laser pulse is executable in a specific scan head location is dependent on the inclination angle of the beam. As stated in Section 4.4.1, the impact of the inclination angle on the crater geometry can be neglected in specific bounds. Therefore the maximum scanning angle of the scan head is constrained (see parameter max_{angle} in the kinematic model). Furthermore a threshold for the maximum deviation of the planned laser pulse direction and the achievable inclination angle in a specific scan head location is introduced.

5.4 Projection into Scan Plane

Determination of the minimum amount of scan head locations necessary in order to process the planned cutting trajectory yields to subsets with laser pulse positions to be executed in the according scan head location. The optimization algorithm determined the number of laser pulses within the working volume of the scan head for each location. Scan heads are usually controlled by providing the coordinates in the local coordinate system. Hence, in a final planning step, it is now necessary to map all three-dimensional laser pulse positions into the local coordinate system of the scan head. The coordinate system of the scan head is two-dimensional and defined in the focal distance of the applied focusing lens. Each coordinate is correlated to specific scanning mirror angles.

Mapping three-dimensional laser pulse positions into the local coordinate system is realized by projection. Figure 5.11 illustrates the approach. Each laser pulse position within the working volume is projected into the focal plane. This projection ensures the correct and accurate three-dimensional placement.

In order to correctly project the three-dimensional laser pulse position into the focal plane of the scan head, the optical center has to be known as well as the location of the focal point (cp. Section 6.3.1). Hence, these

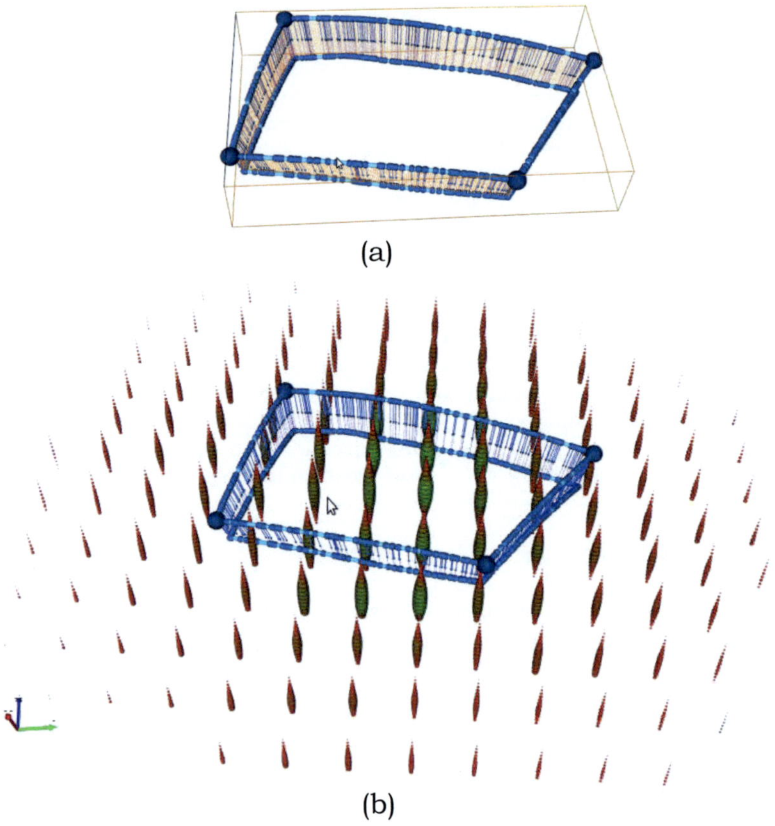

(a)

(b)

Figure 5.10: (a) The oriented bounding box of the cutting trajectory spans the solution space. (b) The bounding box is enlarged and equidistant spacing leads to the family of subsets for evaluation during the optimization. The larger green a possible solution is, the better is the rating. Small red spheres indicate less optimal scan head locations.

two parameters are significant process parameters for the quality of the plan. The laser beam axis between the inclination point and the optical center is used to calculate the intersection point with the scan plane.

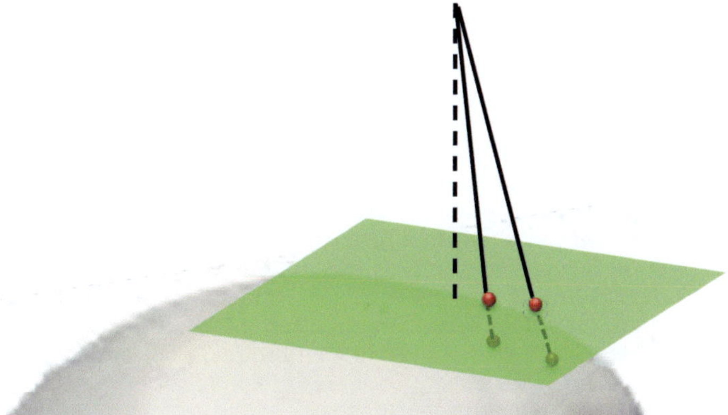

Figure 5.11: Laser pulse positions which are within the working area of the scan head need to be projected into the focal plane in order to determine the corresponding scanning angles.

Projection of all laser pulse position to be processed in a specific scan head location results in a sequence of two-dimensional coordinates. These coordinates are given within scan head coordinate system and thereby directly correlated to the scanning angles. The planning of robot assisted laser bone ablation is finalized with this projection step.

5.5 Conclusion

Robot assisted laser bone ablation necessitates preoperative planning of the laser cut. This chapter introduced the definitions of cutting paths and cutting trajectories on a patient specific surface model. This geometrical definition is then the basis for the determination of laser pulses. The concatenation of these pulses generates the cut. The total amount of laser pulses necessary is strongly dependent on the laser parameters (cp. Chapter 4). Therefore an approximation function was introduced, which estimates the amount of laser pulses required to achieve a specific depth under a predefined pulse duration.

Processing a predefined cut with a two-dimensional scan head necessitates optimal execution planning. A previous optimization method was introduced by Müller [Mül09] in a related diploma thesis. Here, the solution is determined by randomly choosing a subset out of the given laser pulse positions as solution space and evaluating it according to the largest amount of laser pulses covered.

The final optimization methodology for determination of the minimum amount of scan head locations for given ablation pulses is described in Section 5.3. Here, the problem is related to the minimum set cover problem and solved by a greedy algorithm.

Furthermore, laser pulse sequences for a corresponding scan head location are transferred into two-dimensional coordinates in the scan plane. Hence, the robot assisted laser ablation planning leads to scan head locations and corresponding two-dimensional pulse sequences. The developed planning methods are independent of the applied robot. In the next chapter the realized system for robot assisted laser bone ablation is introduced and the developed methods are embedded into the workflow.

6

System Realization for Robot Assisted Laser Osteotomy

In this chapter all necessary components and procedures are described, which constitute the first system for robot assisted laser osteotomy. Applying a laser for the osteotomy of human bone in the operation theater requires a careful system design. This is of particular importance when additionally a robotic system is combined with such a high power laser source. As mentioned in the preliminary studies (cp. Section 3.3.8) the idea of cutting bony tissue using a robot for positioning has been existing for several years. The Institute for Laser Medicine (ILM, Heinrich-Heine University, Düsseldorf, Germany) and the former group for Holography and Laser Technology at the Center of Advanced European Studies and Research (caesar, Bonn, Germany) established the method of short pulsed CO_2 laser bone ablation with assistive water spray and showed feasibility. Unfortunately the research in the scope of laser bone processing could not be continued at these research facilities. Nevertheless some laser hardware components used as part of the system setup in this doctoral thesis are based on the findings published by the mentioned research groups.

The developed system setup for robot assisted laser osteotomy is described in detail in this chapter[1]. After describing the hardware components, the developed registration and calibration procedures of the end-effector are presented. The chapter continues with the description of

---

[1]An intermediate setup was described in [Bur08].

121

the workflow designed for robot assisted laser osteotomy. Furthermore the developed software modules and their interaction are explained.

6.1 System Fundamentals

Following the standardization given in the DIN EN ISO 11145:2008-11 [DIN08] the setup of a laser unit for material processing purposes comprises a laser assembly as well as a manipulation system for relative positioning between laser and the workpiece which is being processed. Optionally a measurement device for controlling the process may be part of the laser unit. Figure 6.1 illustrates the setup of a laser unit following DIN EN ISO 11145. The laser assembly includes the laser device (comprising the laser source and supply), the beam guidance and beam formation.

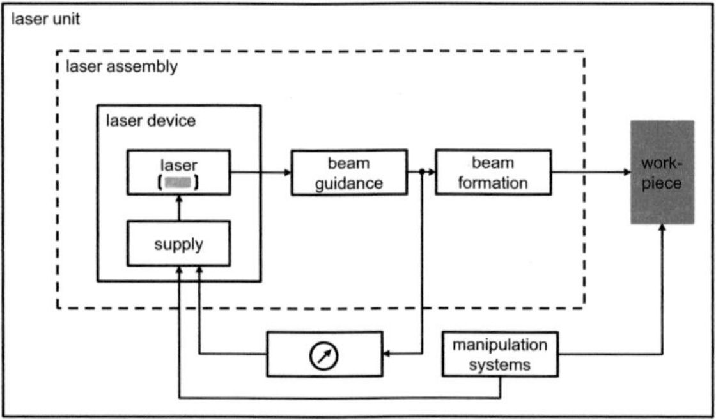

Figure 6.1: Setup of a laser unit following DIN EN ISO 11145 [DIN08].

The idea of positioning the laser beam relative to the bone using a robot is strongly connected to remote laser welding or laser cutting in industrial applications. Hence, the standardized laser unit setup for material processing can be adapted to robot assisted laser osteotomy. Figure 6.2 shows a sketch of the proposed setup. The laser assembly includes the CO_2 laser source and the supply. Beam guidance is realized by an articulated mirror arm suitable for the CO_2 laser, in terms of wavelength and power. The laser beam is thereby delivered to the beam formation, which is a galvanometric two-dimensional scan head. The beam formation unit is positioned by a manipulation system, here

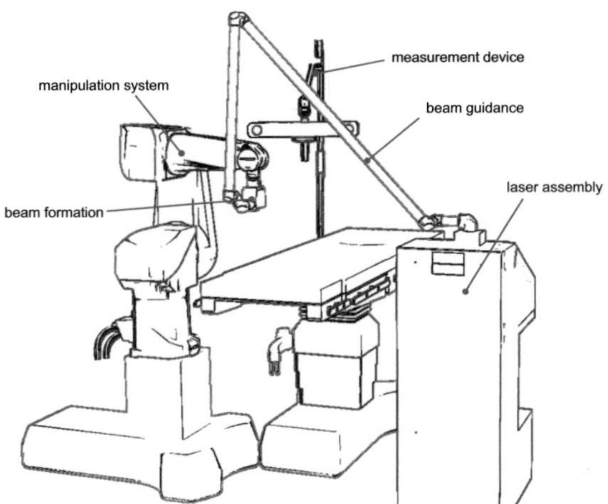

Figure 6.2: Sketch of the system setup for robot assisted laser osteotomy.

a robot manipulator, in order to relatively position the laser beam to the bone. Since the primary aim of this doctoral thesis is to establish a system for robot assisted laser osteotomy and develop a corresponding workflow, the measurement device as proposed optionally by the DIN EN ISO 11145 was not considered in the system setup. However, the current state of the art offers several suitable measurement possibilities, which will be discussed in Section 8.2.

6.2 Hardware Components

In the following the hardware components applied for the established prototype system for robot assisted laser osteotomy will be further described.

6.2.1 Laser System

The prototype laser system OsteoLas x10 was developed by the former group for Holography and Laser Technology at the Center of Advanced European Studies and Research (caesar, Bonn, Germany) [Kla08]. The prototype system was adapted and further enhanced to a second version (OsteoLas x10 v2) in the scope of this doctoral thesis, including optimizations and hardware extensions. The components are embedded into a mobile, compact housing. In the following paragraphs the hardware components comprised by OsteoLas x10 v2 are described.

CO_2 Laser

The CO_2 laser source is a Rofin Sinar SCx10 slab laser (Rofin Sinar Technologies Inc., Plymouth, USA) with a wavelength of $10.6\,\mu m$ and a maximum output power of $100\,W$. This compact laser guarantees an excellent beam profile with $M^2 < 1.2$.

The optical resonator is formed by the front and rear mirrors. Excitation of the laser gas takes place in a radio frequency (RF) field. Between two parallel water-cooled electrode plates the heat generated in the gas is dissipated (diffusion-cooled). A beam shaping module is integrated into the laser head and produces a high quality round symmetrical beam. The technical data is stated in Table 6.1. Since this CO_2 laser is a class 4 laser it is necessary to operate in compliance with the laser security directives.

Table 6.1: Technical Data for Rofin Sinar SCx10 slab CO_2 laser.

Characteristics	Value
Wavelength λ	$10.6\,\mu m$
Beam quality K (M^2)	>0.8 (<1.2)
Maximum Output Power P	$100\,W$
Power Range $P_{min/max}$	$5\,W$ to $100\,W$
Stability (long term)	$\pm 7\%$
Pulse Frequency f	0 - $100\,kHz$
Mode	TEM_{00}

Pilot Laser

With its wavelength of $10.6\,\mu m$, the processing laser is beyond of the visible range for the human eye. In order to allow the user (surgeon) to survey the process optically a red laser diode (Rofin Sinar SC x10 External laser diode option, cp. Table 6.2) is coupled coaxial into the optical axis of the CO_2 laser. Two adjustable mirrors allow the correct collinear overlay of the two beams in the near and far field of the diode laser. The diode laser is a class 3R laser that has to be operated in compliance with the laser security directives.

Table 6.2: Technical Data for Rofin Sinar SCx10 External laser diode.

Characteristics	Value
Wavelength λ	635 nm
Output Power P	4 mW
Beam Diameter	≈ 3 mm

Panel PC

Integrated to the mobile housing of OsteoLas x10 v2 is a panel PC (T15-ERGO, tci GmbH, Heuchelheim, Germany), to control the intervention. The PC is technically built-up following DIN EN 60601 and equipped with a 15 inch display with thermally pre-stressed and coated protection glass. The PC meets the fundamental regulations of the Directive of the Council of the European Community regarding electromagnetic compatibility for clinical environments (89/336/EWG). A touch screen allows the surgeon to input data with gloved hands.

6.2.2 Articulated Mirror Arm

The laser beam is delivered through a passive articulated mirror arm to the galvanometric scan head. Linear arranged aluminum tubes and pivot joints build this passive kinematic. The joints contain reflective mirrors which reflect the laser beam by 90°. The various design possibilities allow to apply an optimized articulated arm for a certain application. Since the laser beam travels a constant distance through the articulated arm, this beam delivery method guarantees a fixed beam size. The quality of an articulated arm is strongly dependent on the calibration and is expressed by the angular and positional accuracy. Furthermore articulated arms arms show constant beam intensity losses what makes them favorable to optical fibers. However, the quality usually declines during the life cycle, since the straight tubes are affected by bending. Furthermore the overall weight of the arm as well as the stiffness of the joints have an impact on the overall accuracy. It is important to notice the complexity of the optical path and account for correct and accurate laser beam coupling.

First experiments were performed with a medical, second hand articulated mirror arm. This mirror arm was designed for hand guidance and therefore the aluminum tubes with a diameter of 26 mm were affected by bending effects. It turned out, that this mirror arm was not suitable for a robot assisted application.

For these reasons an industrial articulated mirror arm which is intended for robot assistance was specified. In the scope of this thesis the specifications and CAD drawings (see Appendix B) were created in collaboration with a manufacturer (Laser Mech Inc., Michigan, USA). Figure 6.3 illustrates the kinematics of the arm and the assigned coordinate systems and the corresponding Denavit Hartenberg (DH) parameters are listed in Table 6.3. Six mirror joints and an additional rotational last joint were chosen in order to allow positioning with 7 degrees of freedom. With an overall weight of ≈ 20 kg, a length of 2231 mm, a tube diameter of 70 mm and stiff joints the arm provides high accuracy, which is of high importance for the application. A spring balancer is utilized in order to keep the arm in an upward position and reduce the affecting weight at the robot's end-effector.

Table 6.3: DH parameters of the mirror arm.

i	a_{i-1}	α_{i-1}	d_i	Θ_i
1	0	$-\pi/2$	0	0
2	0	$\pi/2$	846	0
3	0	$-\pi/2$	0	0
4	0	$\pi/2$	800	0
5	0	$-\pi/2$	86.44	0
6	0	$\pi/2$	0	0
TCP	0	0	211.16	$-\pi/2$

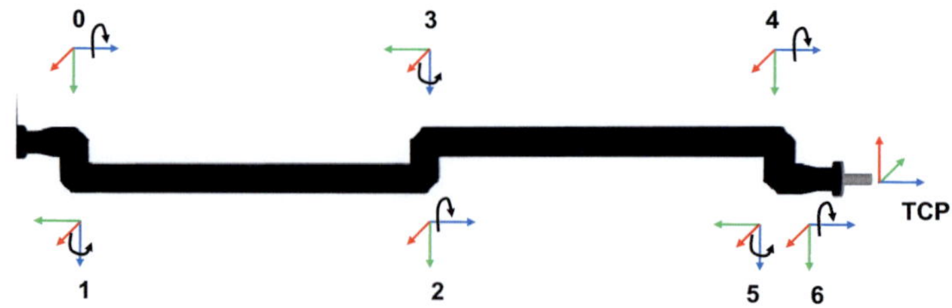

Figure 6.3: Local frames for the articulated mirror arm (following [Cra89]).

6.2.3 Galvanometric Scan Head

In order to distribute the laser pulses accross the tissue quickly, a laser scanning system is utilized. The two-dimensional galvanometric scan head (Colibri, Arges GmbH, Wackersdorf, Germany) has an aperture of 11mm and an accuracy in deflecting the laser of 20 μrad in its working area (70x70mm^2). This scan head is characterized by its compact housing and small weight, thanks to an external galvanometer amplifier and power supply. Figure 6.4 shows the scan head included into an aluminum housing for coupling to the robot and for coupling the articulated mirror arm. The terms *scan head* and *beam deflector* are synonym in the following.

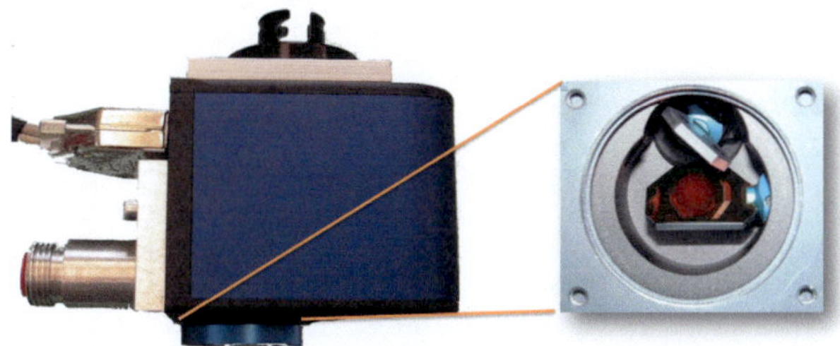

Figure 6.4: Galvanometric scan head from Arges (left). Interior of the scan head (right). Two galvanometric driven coated mirrors deflect the laser beam.

To focus the laser beam a single element ZnSe lens (48TSL100, ULO Optics, Stevenage, United Kingdom) with a specified focal distance of $F = 101.3\,\mathrm{mm} \pm 0.5\%$ is mounted to the scan head. The lens has f-theta characteristics, i.e. the laser beam is focussed onto a plane.

6.2.4 Controlling Laser and Scan Head

The Colibri scan head and the laser are connected to a system controller (ASC2 System Controller, Arges GmbH, Wackersdorf, Germany), which provides real time response to external signals and an optimized synchronization of critical devices. Interfacing the micro controller ASC2 on basis of a Linux OS is realized by TCP/IP; alternatively remote control via RS232 is possible. The Application Programming Interface (API) of the ASC2 controller provided by the manufacturer allows access to all laser parameters. The laser process itself is organized in so called *jobs*, which specify the on-time of the laser in relation to the position in the scan field. The closed-loop control for positioning and acceleration of the galvanometric driven mirrors of the scan head is done by dedicated hardware by high speed digital signal processors (DSP). Actually the API provided by the manufacturer allows to upload jobs, which execution can be started, paused and stopped. There is actually no possibility to have access to the progress of the job processing. While one job is executed, parallel upload of the next one is possible. The abstraction of the laser processing into job files appears as a restriction at the first glance, but since these job files can be designed freely (according to the format), it is possible to define laser processes at the lowest level, i.e. single laser pulses.

6.2.5 Robot

In order to facilitate precise and correct positioning of the galvanometric scan head in respect to the bone, a robot system is necessary. The developed methods for robot assisted laser osteotomy are independent of the specific robot model. The medical application (workspace) and the accuracy demands provide a statement of requirements to the robot kinematics and characteristics. In the scope of this thesis a Stäubli RX90B CR (Stäubli Tec Systems GmbH, Bayreuth, Germany) robot was used. Additionally a light-weight redundant robot kinematic (KUKA LWR, Kuka Roboter GmbH, Augsburg, Germany) was evaluated in a simulation and first experiments were carried out.

The main research work was performed using the Stäubli RX90B CR robot. The RX90 is a conventional industrial serial robot with six degrees of freedom, maximum payload of 12 kg and a maximum reach of 900 mm. With its high repeatability of ±0.02 mm, high stiffness and large workspace, this robot is optimal for precise positioning of the galvanometric scan head. The model utilized here is a clean room (CR) robot, which was formerly used as the CASPAR system from ortoMaquet (cp. Section 3.2.5). However, the robot has a relatively large footprint and is therefore unfavourable for use in an operation theatre. Figure 6.5(a) and (b) illustrate the setup with this robot. See Appendix C.1 for specification details.

The Stäubli RX90B robot comes along with a CS7 controller with the real-time and multi-tasking operating system V+ from Adept technologies (Adept MV19 controller, V+ programming language Version 12.3). V+ manages all system level operations, such as IO, program execution and task management. Communication is realized via serial port RS232. There is no API provided. A self written program parses incoming commands. Robot movements are commanded in joint space.

Beside, first experiments were performed using the LWR from Kuka. With a weight of only 15 kg this robot is light weight (weight of Stäubli RX90B is 111 kg), allowing a payload of maximum 14 kg according to the specification of the manufacturer. The maximum range is 868 mm. Furthermore the kinematics shows reduncancy with a seventh joint allowing null space movement. The lower stiffness of this robot introduces less accuracy in comparison to conventional industrial robots. Therefore additional sensory is necessary in order to improve the accuracy. With an optical tracking system this improvement can be realized by applying methods of visual servoing [Moe09]. Figure 6.5(c) and (d)

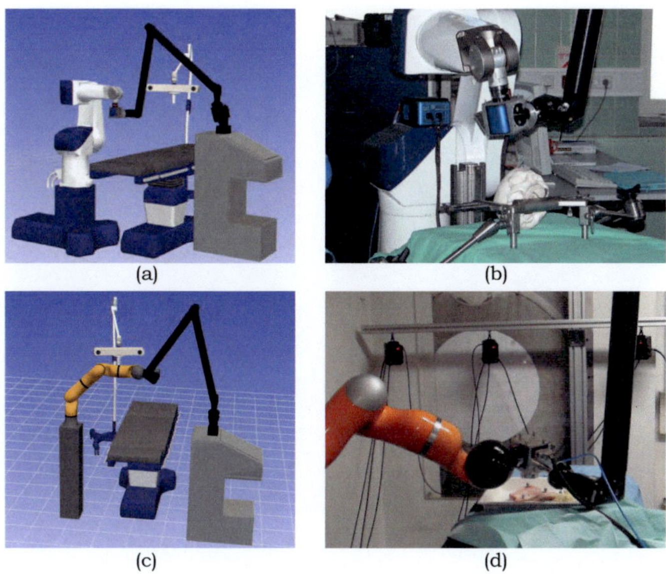

(a)   (b)
(c)   (d)

Figure 6.5: Two types of robots are utilized for robot assisted laser osteotomy. (a) and (b) show the setup using a Stäubli RX90B CR robot. (c) and (d) depict the setup with a Kuka lightweight robot.

illustrate the setup with the light weight robot. See Appendix C.2 for specification details.

6.2.6 Experimental Setup

Figure 6.6 shows the experimental setup for robot assisted laser osteotomy as it was realized in the scope of this doctoral thesis. In the following the system components are summarized. The Stäubli RX90B CR robot positions the scan head relatively to the specimen. The scan head is mounted to the robots flange using a manual gripper changing system (MGW50, Grip GmbH, Dortmund, Germany). The laser is situated inside the mobile housing with the panel PC. The articulated arm deflects the laser beam to the scan head. A spring balancer under the ceiling keeps the articulated arm in an upward position and reduces the load for the robot.

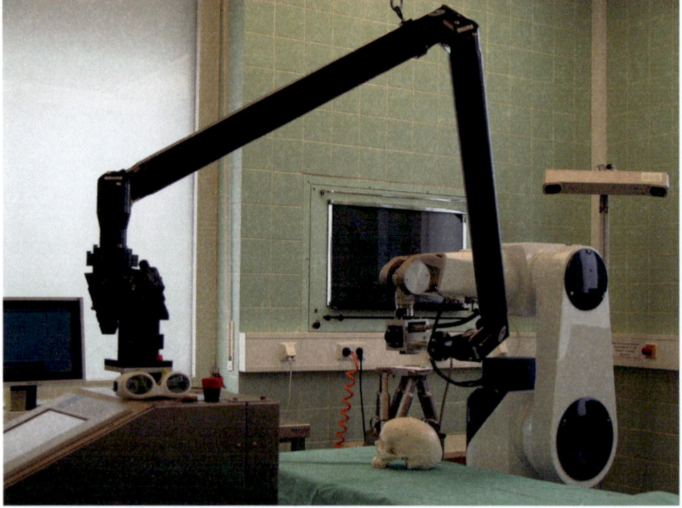

Figure 6.6: The experimental setup for robot assisted laser osteotomy.

6.3 Calibration and Registration

The established system setup for robot assisted laser osteotomy comprises several components and hence inherits several coordinate systems. Figure 6.7 illustrates the relations.

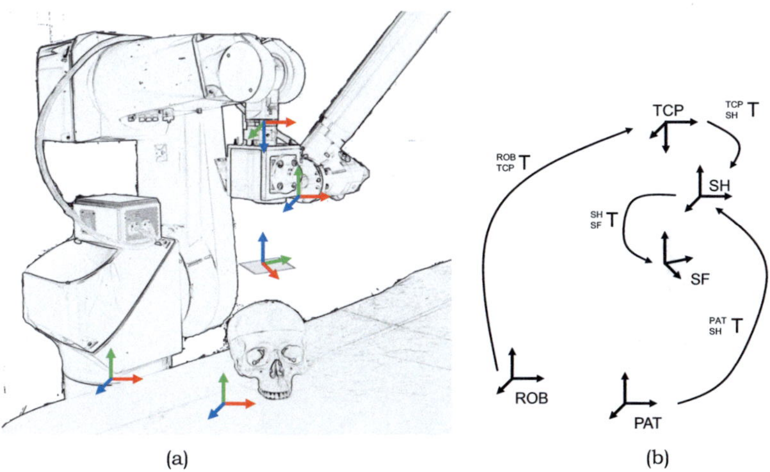

Figure 6.7: (a) Involved coordinate systems. (b) Homogeneous transformation matrices between coordinate systems.

The transformation between the robots tool center point coordinate system TCP and the scan field coordinate system SF is composed of two sub-transformation matrices $^{TCP}_{SH}T$ and $^{SH}_{SF}T$, which are determined in the *calibration and registration of the end-effector* [Bur09d]. Subdividing the transformation determination facilitates interchangeability, e.g. the use of any robot or scan head (manufacturer, lenses). The transformation matrix $^{PAT}_{SH}T$ is related to the *patient registration*, which determines the correspondence between the planning coordinate system (i.e. the CT coordinate system) and the intraoperative situation.

In the following the developed calibration and registration procedures for the end-effector are introduced. Afterwards the patient registration is summarized.

6.3.1 Calibration and Registration between Scan Head and Scan Field

Registration of the scan head coordinate system relative to the scan field coordinate system implies the determination of the homogeneous transformation matrix $_{SF}^{SH}T$ between the two coordinate systems (see Figure 6.8). The registration and calibration of the scan head to the scan field coordinate system necessitates four steps:

- Perpendicular beam adjustment

- Determination of the optical center

- Correction of the scan field distortion

- Determination of the focal distance

- Scan field registration

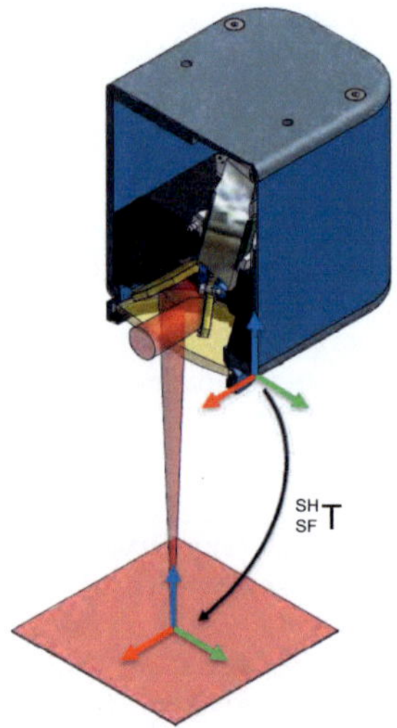

Figure 6.8: The transformation $_{SF}^{SH}T$ between the coordinate system of the scan head and the coordinate system of the scan field have to be determined. (Scan head drawing by courtesy of Arges GmbH, Wackersdorf, Germany)

Perpendicular Beam Adjustment

Under the assumption that the laser beam is coupled into the beam deflector perfectly collinear and a proper calibrated beam deflector, the laser beam is supposed to be perpendicular to the scan plane in the zero position. As an essential parameter for the planning of laser ablation this has to be proven regularly. In order to prove if the laser beam is perpendicular to the scan field, a paper sheet is adjusted parallel to the beam deflector underside using a hexapod. A square ablation pattern and the center point is processed in varying distances to the beam deflector underside. Figure 6.9 illustrates the effect of an angular error of the laser beam compared to a perfectly perpendicular beam output. Iterative adjustment of the x- and y- galvanometer offsets in angular degree is necessary in order to correct the angular error in the beam output. In the scope of this thesis this method was applied for correcting an angular error of $1.8°$.

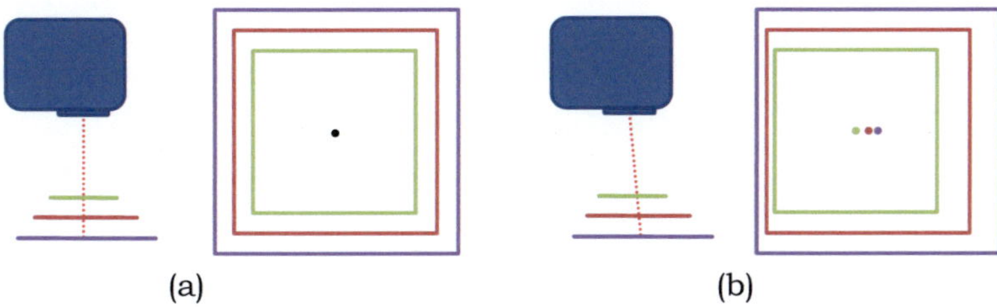

(a)   (b)

Figure 6.9: (a) A perpendicular laser beam results in centric aligned squares processed in varying distances. (b) An angle error results in squares out of alignment.

Determination of the Optical Center

The applied flat field focusing lens results in a scan field which is planar (cp. Section 4.3.4). In order to apply laser pulses onto surfaces with a curvature it is essential to project the inclination point into the scan field (cp. Section 5.4). Hence, the optical center has to be determined. Here the experimental setup described in the previous section can also be applied. Adjusting a sheet of paper parallel to the beam deflector underside in two different distances and processing of a square ablation pattern in each distance can be used to calculate the optical center location by the theorem of intersecting lines.

Correction of the Field Distortion

To compensate for the field distortion caused by the focusing optic and non-linear behavior of the mirror galvanometers of the scan head, the corrective method provided by the manufacturer is utilized. The target material (thermal paper) was positioned in parallel to the scan head with in the focal distance on a hexapod and a test pattern was applied with 20 µs pulse duration and a repetition rate of 200 Hz. Measuring distances between the marks allows the calculation of the scan field correction parameters and leads to a deskewed scan field at the focal distance (cp. Section 4.3.4, Figure 4.12). The corrected scan field is a square of sides 70 mm.

Experimental Determination of the Focal Position

Theoretically the employed single element ZnSe lens (48TSL100, ULO Optics, Stevenage, United Kingdom) has the focal distance F = 101.3 mm ±0.5%. To ensure optimal cutting results, the focal position has to be identified experimentally. From the numerous different techniques for establishing the focus position of a lens, the drilling test is utilized [Pow98, Kla83].

Under the assumption that the smallest beam diameter (i.e. at the focus) will drill the smallest hole, single laser pulses with a pulse duration of 100 µs are applied in varying distances to acrylic material. The chosen distances are approximately 12 mm further than the quoted manufacturers distance, due to the mounting of the lens to the scan head. The experiment is repeated using pulse durations of 30 µs. The acrylic sheets had a thickness of 8 mm. The low thermal conductivity of acrylic sheet is highly absorptive to the laser beam. The target material is adjusted in parallel to the bottom side of the scan head using a hexapod (see Figure 6.10). Acquiring points on the hexapod and on the underside of the scan head with a measurement arm allows to adjust the hexapod in parallel with varying distances. After applying one laser pulse, the hexapod moves 0.25 mm away from the scan head underside and laterally 1 mm in the plane. Figure 6.11(a) shows a microscopic view of the acrylic sheet after applying all laser pulses. Figure 6.11(b) illustrates the laser beam waist. At the focal point the diameter of the drilled hole is the smallest.

The resulting holes are analyzed using a digital microscope (VHX-600, Keyence) with 200x zoom. To apply a quantitative value which reflects this observation, all pixels belonging to the drilled hole are segmented by a manually determined threshold in each picture. The number of

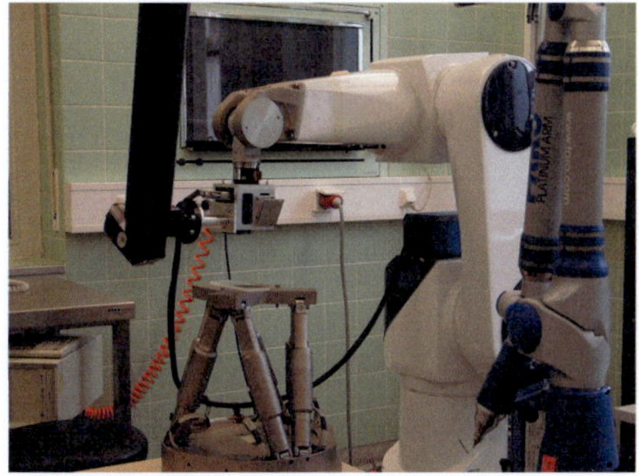

Figure 6.10: A hexapod is utilized to position the acrylic sheet accuratly parallel and in varying distances to the scan head.

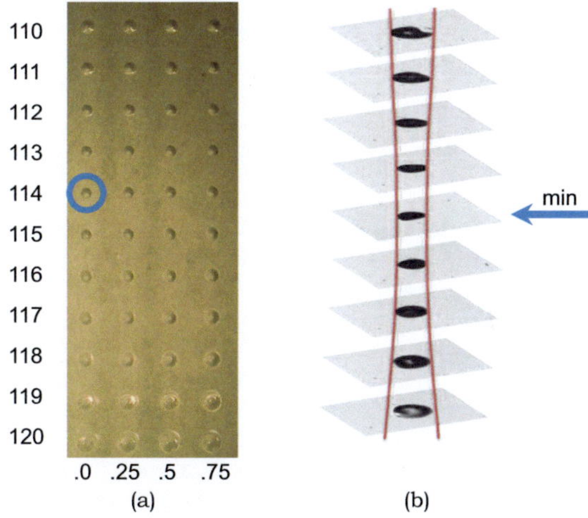

Figure 6.11: (a) Overview of the acrylic sheet after applying single laser pulses in varying distances from the scan head (all values in mm). Marked in blue is the pulse at 114 mm, where the crater is circular and with smallest diameter. (b) Illustration of the focused laser beam: The beam waist is corresponding to the focal distance.

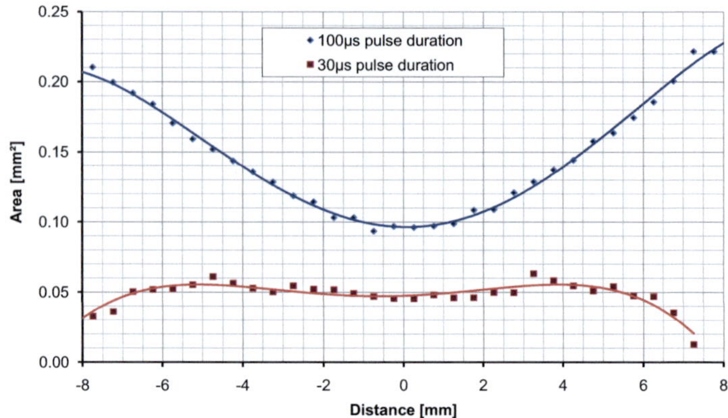

Figure 6.12: The ablation area (in mm^2) is plotted against the distance (in mm) of the scan head to the target material. The regression function (4th degree polynomial) has its low point at the focal distance.

pixels is correlated to the area of the hole (scaling: $1\,\text{mm}^2 \mathrel{\hat{=}} 386884$ pixels). In Figure 6.12 the ablation area (mm^2) is plotted against the distance (mm) of the scan head to the target material. Applying a regression function (4th degree polynomial) and calculating the low point leads to the focal distance, which corresponds to the z value in the translational part of the transformation matrix $^{SH}_{SF}T$.

The area removed by 30 µs pulses is significantly smaller due to the lower intensity applied to the material. The two regression functions have their minimum at the same point, which is the focal distance F = 114 mm measured from the underside of the scan head. As stated in Section 4.3.5, the focal distance of the lens (F = 101.3 mm) is slightly shifted farther for the applied optical path about 1.4 mm. The missing summand results mainly from the distance between the scan head underside and the principal plane of the mounted lens.

Registration of the Scan Field

To determine the remaining values of the transformation matrix $^{SH}_{SF}T$, the x- and y-axis of the scan field need to be described relatively to the scan head coordinate system. A sheet of thermal paper is mounted

136

onto a hexapod. By measuring three points on the scan head underside and three points on the hexapod platform the hexapod moves relatively to the scan head in a parallel position at the focal distance. In this position a pattern of single laser pulses is applied with a pulse duration of 30 µs. The pattern corresponds to the two-dimensional coordinate system of the scan field, defined by the manufacturer. The laser pulses are distributed with 2.5, 5, 10, 15 and 20 mm from the origin along the x- and y-axis (see Figure 6.13).

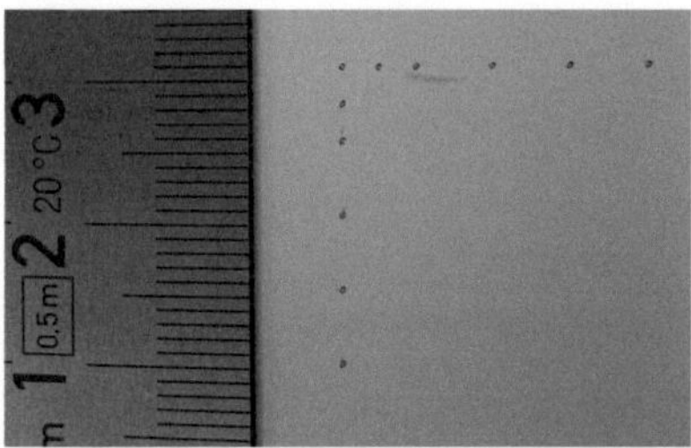

Figure 6.13: After adjusting a sheet of thermal paper in the focal distance parallel to the scan head using a hexapod, a pattern of single laser pulses is applied for registration.

The laser pulse incisions are measured using a Faro Platinum Arm (FARO Technologies, Inc., USA) with a point probe. Additionally a surface scan of the scan head is taken using the laser scanner of the Faro Platinum Arm. The acquired laser pulse incisions and the point cloud are shown in Figure 6.14a together with the CAD model of the scan head before registration. The two surfaces of the scan and the CAD model are matched using the Iterative Closest Point (ICP) algorithm. The resulting transformation matrix is used to transform the measured laser pulse incisions into the coordinate system of the scan head. Figure 6.14b shows the matched surfaces and the measurement of the incisions in thermal paper.

With the coordinates of the laser pulses in the scan field coordinate system and in the scan head coordinate system, the closed form solution of the orientation problem using unit quaternions proposed by Horn [Hor87] is used to determine the transformation matrix. The resulting matrix comprises approximately an identity matrix for the rotational part and a translation. The identity rotation matrix approves the perpendicular beam adjustment.

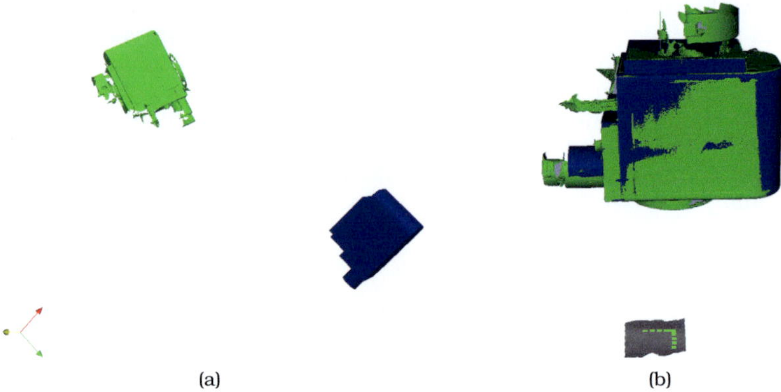

(a)   (b)

Figure 6.14: (a) Before Registration. In the upper left corner the measured point cloud of the beam deflector (dark green) and the ablation pattern (light green) in the coordinate system of the measurement arm. In the lower right corner the CAD model of the beam deflector. (b) After successful surface matching. The measured points are now transformed into the coordinate system of the beam deflector.

The resulting matrix of the registration of the scan field in scan head coordinates is

$$
{}^{SH}_{SF}T \;=\; \begin{bmatrix} 1 & 0 & 0 & -35.75 \\ 0 & 1 & 0 & -38.92 \\ 0 & 0 & 1 & -114 \\ 0 & 0 & 0 & 1 \end{bmatrix}
$$

This point based registration has a fiducial registration error (FRE) [Fit98] of 0.21 mm, which is in the same range as the diameter of the laser beam ($^1/e^2$) in its focal point ($\approx 200\,\mu m$).

In order to simplify the approach for continuous verification of the transformation ${}^{SH}_{SF}T$, defined landmarks on the beam deflector housing are used instead of conducting a laser scan every time. However, the laser scan is independent from the adapter or housing of the beam deflector and therefore more general.

6.3.2 Registration between the Tool Center Point and the Scan Head Coordinate System

The second part of the end-effector registration is the determination of the transformation matrix between the flange coordinate system of the robot and the beam deflector coordinate system $^{TCP}_{SH}T$. This transformation matrix can be derived from the CAD drawings of the scan head with the mounted adapter plates, the manual gripper changing system and the robot itself. In order to validate the transformation matrix, the flange of the robot and the mounted scan head are measured with a Faro Platinum Arm. Figure 6.15 illustrates the measurement points. First, three points at the flange underside are measured in order to determine the x-/y-plane and three points are taken of the circular flange

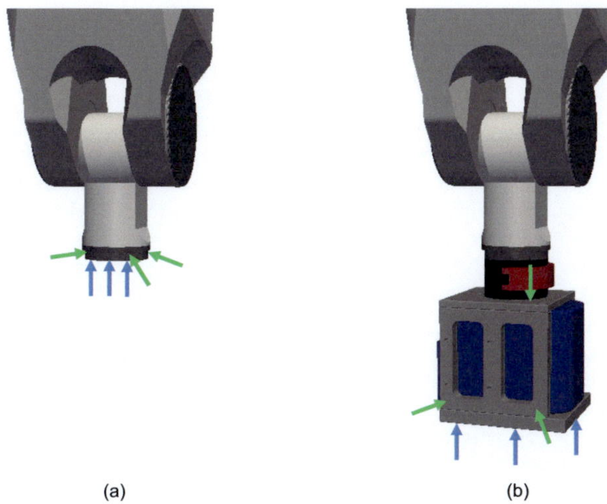

(a)   (b)

Figure 6.15: For the end-effector registration the transformation between the flange coordinate system of the robot and the beam deflector coordinate system need to be determined. (a) Measurement points at the flange: Three at the circle (green arrows) and three of the x-/y-plane (blue arrows). (b) Measurement points at the mounted scan head housing: Three registration points (green arrows) and three at the underside (blue arrows).

to determine the z-axis. Second, the scan head is mounted to the flange and registration points are measured. Additionally three measurement points are taken randomly at the bottom side of the scan head in order to determine an possibly encountered tilt between the flange x-/y-plane. The resulting matrix of the registration of the tool center point (TCP) is

$$
{}^{SH}_{TCP}T \;=\; \begin{bmatrix} -1 & 0 & 0 & -48.81 \\ 0 & 1 & 0 & -32.37 \\ 0 & 0 & -1 & 86.10 \\ 0 & 0 & 0 & 1 \end{bmatrix}
$$

6.3.3 Patient Registration

In order to execute a preplanned cutting trajectory with the robot assisted laser osteotomy system the patient has to be registered, i.e. the transformation matrix between the planning coordinate system and the actual patient coordinate system has to be determined ${}^{CT}_{PAT}T$. This transformation is calculated by utilizing the gold standard method: Point-based registration (cp. 3.1.4). Hence, the target bone is equipped with titanium marker screws before image data acquisition. The locations of the screws (fiducials) are determined in the CT datasets´. They define the planning coordinate system. Before starting the execution of the cutting trajectory intraoperatively, the fiducials are localized utilizing a measurement arm. By applying point-based registration the transformation matrix ${}^{CT}_{PAT}T$ can be determined.

However, beside the knowledge of the transformation of preplanned scan head locations from the CT coordinate system to the actual patient location it is necessary to know the spatial relation of the robot system to the patient (cp. Figure 6.7b). In the scope of this thesis the transformation matrix ${}^{PAT}_{SH}T$ is therefore measured in the patient registration step. The robot moves into a defined so-called *registration pose*. Hence, the transformation ${}^{ROB}_{TCP}T$ is known. Furthermore the transformation matrix ${}^{SH}_{TCP}T$ is known from the registration procedure introduced in the previous section (cp. Section 6.3.2). By measuring three well defined points on the scan head the spatial relation between the patient coordinate system and the scan head coordinate system can be determined. However, the transformation ${}^{PAT}_{SH}T$ is only valid for the registration pose of the robot. Therefore the transformation ${}^{PAT}_{ROB}T$ is calculated by

$$
{}^{PAT}_{ROB}T = {}^{PAT}_{SH}T \cdot {}^{SH}_{TCP}T \cdot {}^{ROB}_{TCP}T^{-1}.
$$

The transformation between the robots base coordinate system ROB and the patient coordinate system PAT remains static as long as the patient does not change his location. Since no online tracking of the patient location is performed in the scope of this doctoral thesis, the patient is supposed to be fixated in respect to the robot.

6.4 Workflow

Performing an osteotomy using robot assisted laser ablation in the operation theatre requires an appropriate workflow. In the scope of this thesis the workflow illustrated in Figure 6.16 was developed.

The first step is the acquisition of a CT dataset of the patient. Afterwards a surface model is derived from segmenting the bony structures in the datasets and applying the marching cubes algorithm for determination of the triangulation. This surface model is used for planning the geometry of the incision. This geometrical incision definition needs

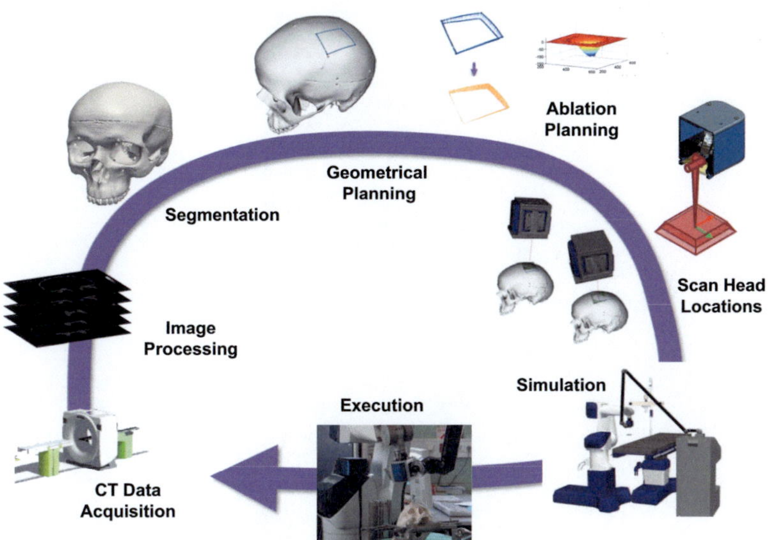

Figure 6.16: Symbolic overview of the workflow for robot assisted laser processing of bone.

to be transferred into an ablation pattern, which is executable by the laser system afterwards. Ablation planning therefore utilizes the discrete laser process model in order to derive a prediction for the amount of laser pulses and their position. Based on this ablation pattern locations for the scan head are determined for optimal execution planning. In a simulation step the execution of the laser cut is evaluated, before the real execution can take place.

All steps of the workflow are realized in the scope of this thesis. Starting from the planning phase new, innovative methods are developed and investigated. The most important steps are highlighted in Figure 6.17.

during the *Planning Phase* the user defines the geometry of the osteotomy on the surface model. This geometrical definition has to be transferred into a corresponding ablation pattern, since laser bone ab-

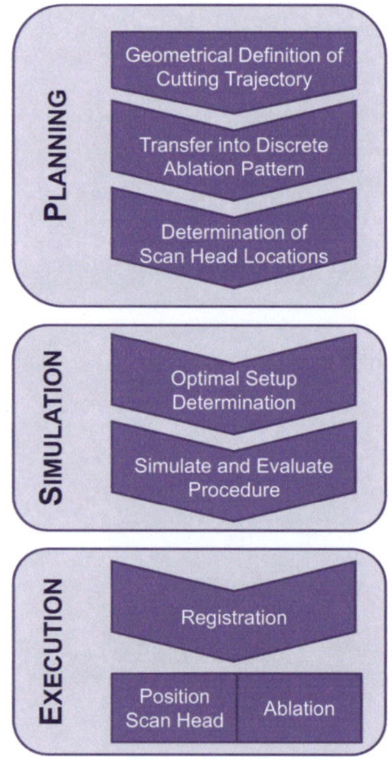

Figure 6.17: Workflow of the planning, simulation and execution phase for robot assisted laser osteotomy.

lation is a discrete process composed of multiple single laser pulses, each removing a tiny bone piece, concatenated to the complete cut (cp. Chapter 4). Based on this pulse distribution the execution with the robot and scan head has to be planned. Hence, the optimal amount of scan head locations suitable to process the ablation pattern has to be determined (cp. Section 5.3). The scan head deflects the laser beam in its working area in the desired ablation pattern corresponding to the section of the cutting trajectory to be processed in the specific scan head location.

In order to evaluate the resulting scan head locations and ablation pattern, a *Simulation Environment* was developed. This offers the possibility for determining the optimal setup of all components. On the one hand manual setup planning is supported. On the other hand the simulator allows to integrate the registration information of the patient and check feasibility for this case. Furthermore the Simulation Environment facilitates to visualize the ablation procedure in advance.

After successful planning and simulation the *Execution Phase* takes place, i.e. the intraoperative phase. The patient is fixated to prevent any movement during the laser ablation process. For registration of the patient the gold standard method with titanium marker screws implanted preoperatively before acquiring image data is utilized. The registration is determined according to Section 6.3.3. After successful registration the robot moves into the first scan head location, where it remains until the corresponding ablation pattern was executed by the scan head. Iteratively moving through all scan head locations the cutting trajectory is processed, ablating layer by layer the bone following the preoperative plan.

6.5 Software Modules

In order to implement the described workflow some software modules are necessary. These are the Planning Module, the Simulation Module and the Execution Module. The established process model and the parameters of Chapter 4, as well as the methods for planning and optimization introduced in Chapter 5, are consolidated in these modules. The functionalities are described in the following subsections.

6.5.1 Planning Module

For preoperative planning of robot assisted laser bone ablation a planning module was developed in the scope of this doctoral thesis. Implementation is based on the open source libraries vtk (Visualization ToolKit, Kitware Inc. Clifton Park, New York, USA) and Qt (Qt Development Frameworks, Nokia, Oslo, Norway). The planning module comprises the geometry planning and the ablation planning, both explained in the following:

Geometry Planning Module

The geometry planning module realizes the methods introduced in Section 5.1. The user defines support points on a patient specific surface model of the target bone. The waypoints are generated automatically. The module allows definition of single locations, marking paths as well as cutting trajectories, where the user additionally specifies the cutting depth and cutting angle. Figure 6.18 illustrates planning of a cutting geometry.

Ablation Planning Module

In order to determine a distribution of single laser pulses to be executed by the laser system for achieving the predefined cut the ablation planning modules implements the methods described in Section 5.2. The transfer of geometrical cutting information into a corresponding sequence of laser pulses is performed automatically. The user predefines the laser parameters (e.g. repetition rate, pulse duration). Figure 6.19 depicts an exemplary transfer of an incision into an ablation sequence.

Based on the generated laser pulse sequence execution planning is performed (cp. Section 5.3). A modified greedy algorithm is implemented for automatically determining the minimal amount of scan head locations for execution of the laser cut. Figure 6.20 illustrates exemplary scan head locations for a laser cut.

The overall amount of laser pulses necessary to perform the predefined cut are partitioned according to the scan head location, i.e. all laser pulses not covered in a specific scan head location are discarded. Furthermore all laser pulse positions are projected into the focal plane of the scan head along the laser axis (i.e. the line connecting the laser pulse position and the the position of the optical center). Thereby the two-dimensional coordinates of the necessary laser pulses in the coordinate system of the scan head are determined. For each scan head location the resulting scan head job is stored for execution.

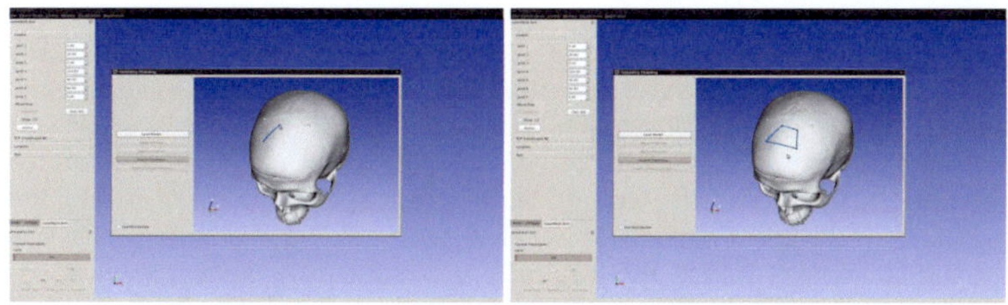

Figure 6.18: The geometry planning module allows the user to define arbitrary cuts on the patient specific surface model by setting support points.

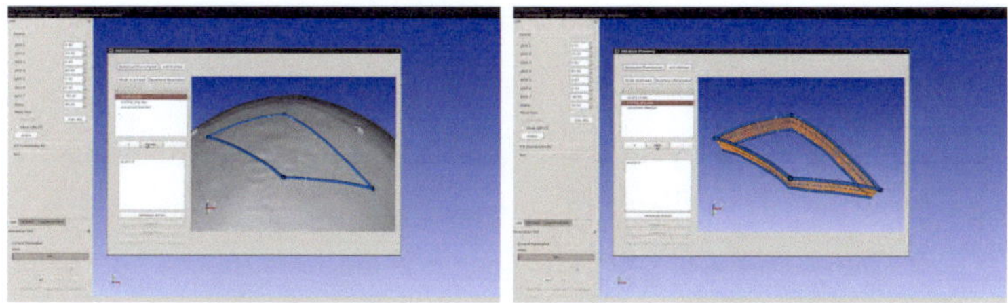

Figure 6.19: The geometrical definition of the cut is automatically transferred into a corresponding ablation sequence.

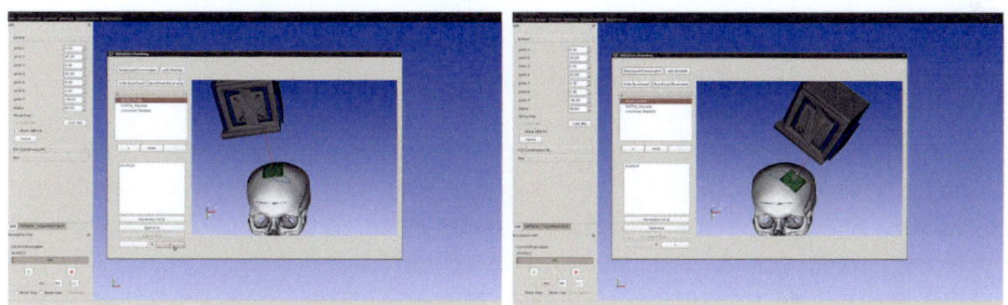

Figure 6.20: Scan head locations are determined automatically for a laser cut.

6.5.2 Simulation Module

The simulation environment (see Figure 6.21) provides a preoperative and intraoperative patient individual simulation for robot assisted laser bone ablation. It is important to notice, that the generated scan head locations are independent of the robot which will be utilized. The simulation offers the possibility to check, whether a chosen robot model with an articulated mirror arm is suitable to perform the generated cut. Configuring the intervention by choosing a robot and a mirror arm model, necessitates also consideration of the setup also. Evaluation of the combined workspace is the first step to prove feasibility of the planned laser osteotomy. By manual setup of the components and integration of registration information (i.e. placement of the patient relative to the robot) the intervention can be simulated in advance.

Implemented simulation functionalities are listed in the following:

- 3d visualization of planned trajectory and ablation patterns,

- Multi-body simulation of active and passive kinematic chains,

- Simulation of different scenarios by varying robot model and articulated mirror arm model,

- Feasibility study of different specifications for robots and articulated mirror arms,

- Feasibility study of intervention: Planned end-effector locations reachable and collision free execution?,

- Verification and validation of planning data,

- Simulation of robot assisted laser osteotomy intervention (e.g. robot motions, placement of single laser pulses).

All models represented in the virtual scene are necessarily involved in the surgical procedure, i.e. namely the patient's skull (anatomy), the robot, the laser system, the mirror arm, scan head, sensors and operating table used. All 3d models of the hardware components are derived from CAD drawings of the manufacturer and represented in StereoLithography (stl) format.

Since beam delivery by an articulated mirror arm is an important issue in robot assisted laser osteotomy the simulation environment offers a specification module for passive kinematic chains. The user can easily define the number of joints and link lengths of the kinematic chain. Forward and backward kinematic equations are automatically generated using Orocos Kinematic and Dynamics Library (KDL) [Oro09]. Thereby movements indicated by the robot can be simulated and the combined workspace can be visualized.

After preoperative simulation the final planning data necessary for the intraoperative execution is generated, which now also includes information on the used components and optionally their setup in the OR (positions of robot, laser and patient relative to each other). If no setup information is included, the perioperative registration informa-

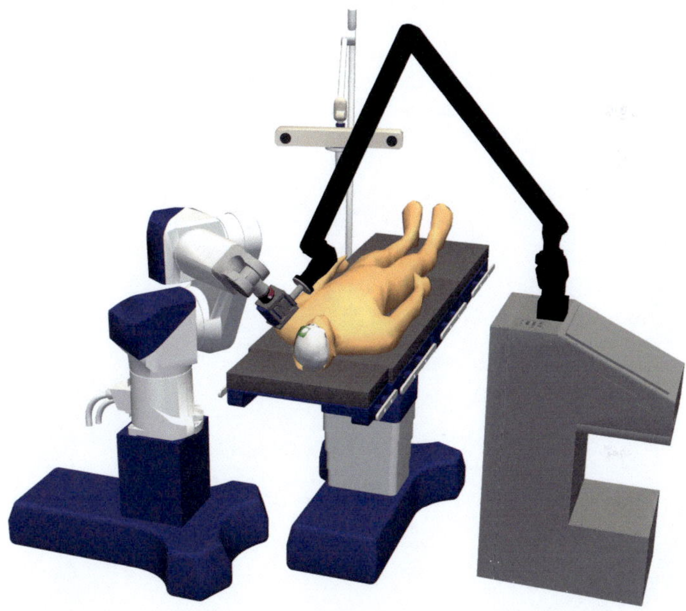

Figure 6.21: Simulation of executing laser osteotomy on a human skull for craniotomy. The laser is delivered via the articulated mirror arm to the scan head. The scan head is attached to the robots flange, here a Stäubli RX90B CR robot. An optical tracking system is utilized for online location measurement, here the NDI Polaris system.

tion needs to be loaded into the simulation and the feasibility of the intervention has to be checked during the perioperative phase, possibly necessitating relocation of the patient or the devices.

6.5.3 Execution Module

The preoperative planning module provides the control sequences for executing robot assisted laser osteotomy, i.e. patient registration information, scan head locations and corresponding scan head jobs. The Intervention Control software interprets the planning data and is then in charge of execution and supervision of the intervention. Hence, the software orchestrates the robot, the laser and the scan head, as well as a localization system (e.g. optical tracking system or measurement arms). Furthermore it is already foreseen to integrate additional sensory for online observation of the laser ablation process. Figure 6.22 illustrates the intervention control.

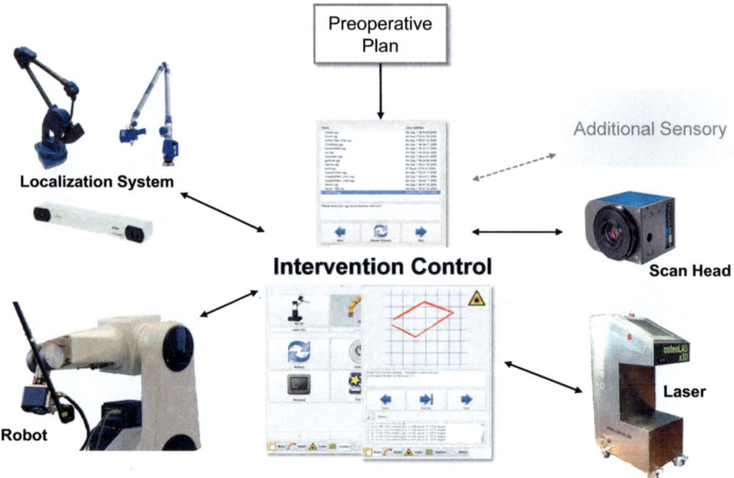

Figure 6.22: The Intervention Control orchestrates the hardware components. The user is guided step by step through the intervention by the GUI and is in charge of activating the next step.

Interpretation of the planning data results in a finite state machine. Since a surgical intervention is not considered to be automatable completely, the user, i.e. the surgeon is the highest authority deciding when state transitions take place. The graphical user interface is run by the Intervention Control on the panel PC (see Section 6.2.1) to guide the user (surgeon or OP personnel) through each step by a well-structured and clearly presented visual representation of each state of the finite state machine. Typical tasks are described to the user in a more colloquial medical or technical language.

6.6 Conclusion

The system realized for robot assisted laser osteotomy developed in the scope of this doctoral thesis was introduced in this chapter. It represents world's first realization of a complete system for laser bone processing. The system was designed according to DIN EN ISO 11145:2008-11, comprising a laser assembly, an articulated mirror arm for beam guidance, a two-dimensional scan head for beam formation and a robot as manipulation system.

Laser bone ablation requires an accurate system calibration in order to facilitate the high cutting precision obtainable by the laser process. Therefore adequate calibration and registration methods were developed in the scope of this doctoral thesis. Furthermore the workflow for robot assisted laser osteotomy comprising preoperative and intraoperative process steps was introduced in this chapter. In the scope of this doctoral thesis all workflow steps were reconsidered. Planning methods for cutting bony tissue with laser and the transfer into executable sequences of laser pulses were implemented in corresponding software modules. A simulation step before the execution allows to verify feasibility of the plan and optimize the system setup in the operating room.

The complete realization and development process for the system for robot assisted laser osteotomy was accompanied by a risk analysis according to DIN EN ISO 14971. The identification of reasonable hazards and their causes under normal and fault conditions affected the development decisions towards risk reduction. In total 200 risks were identified and evaluated. In order to prevent or at least to reduce these risks, risk control measures are proposed and evaluated according to their efficiency. Some measures are already implemented, e.g. the graphical user interface (GUI) of the intervention control was revised substantially for fail-safe user guidance. In Annex A the assessed risk analysis is summarized.

7
Experimental Feasibility Studies

For the first time an overall system for robot assisted laser osteotomy and a corresponding workflow have been established. In the state of the art the high potential was already anticipated by different authors, but an overall experimental verification of this hypothesis can be performed for the first time in the scope of this doctoral thesis.

To evaluate the overall accuracy of the established system for robot assisted laser osteotomy a series of experimental feasibility studies was performed. The first experimental series dealt with marking trials on a human skull replica. The complete workflow was accomplished, starting with image data acquisition and surface model determination. Pre-definition of the geometry of the marking paths and transfer into an optimized execution plan were performed prior to the experiment. The experiments themselves include the registration of the skull and execution of the laser cuts from multiple robot locations. Post-experimental evaluation reveals the system capabilities.

Afterwards the accuracy and the system performance were evaluated on ex-vivo animal bone specimen. The cutting experiments were accomplished in exact the same manner as the marking experiments. Performed osteotomies are analyzed in post-experimental CT datasets.

Beside the determination of the overall system accuracy, a medical application was regarded in the scope of the experimental feasibility studies. An experimental setup trial was conducted in order to show OR feasibility of the developed system in the scope of robot assisted laser based cochleostomy.

151

7.1 Marking Experiments

In order to evaluate the system capabilities the first experimental series dealt with marking trials performed on a human skull replica. In the case of marking a few consecutive layers of single laser pulses are planned according to a marking path.

7.1.1 Preparation

A human skull replica (A20, 3B Scientific GmbH, Hamburg, Germany) made of hard plastic was equipped with four titanium marker screws for registration. A computer tomography (CT) scan with a slice distance of 0.6 mm and x- and y-resolution of 0.39 mm was acquired. Utilizing the imaging software and DICOM viewer OsiriX the CT dataset was postprocessed. A surface model was derived by segmenting the skull *bone* and afterwards applying the marching cubes algorithm for the determination of a triangulated mesh.

7.1.2 Planning

Marking paths were defined on the three-dimensional surface model by defining support points. Figure 7.1 illustrates the planned markings paths. In the following the trajectories will be referred to as line, triangle and cuboid according to their shape. The geometrical plans ware transferred into ablation patterns. If one working volume per trajectory is not sufficient the path is divided into several scan head locations.

Based on the ablation pattern optimized scan head locations were automatically determined following Section 5.3. The optimization results in one scan head location for the triangle, two scan head locations in order to cover the cuboid and the rectangle as well as three scan head locations for the line (cp. Figure 7.2). Figure 7.3 additionally illustrates the relative position of the three scan head locations for the line.

7.1.3 Execution

For the experiments the skull was fixated in a Mayfield clamp like device and repositioned after execution of one marking path. For registration a Faro Platinum Measurement Arm equipped with a point probe was utilized to measure the location of the fiducials and to determine the relative location of the skull to the robot. The registrations had a FRE

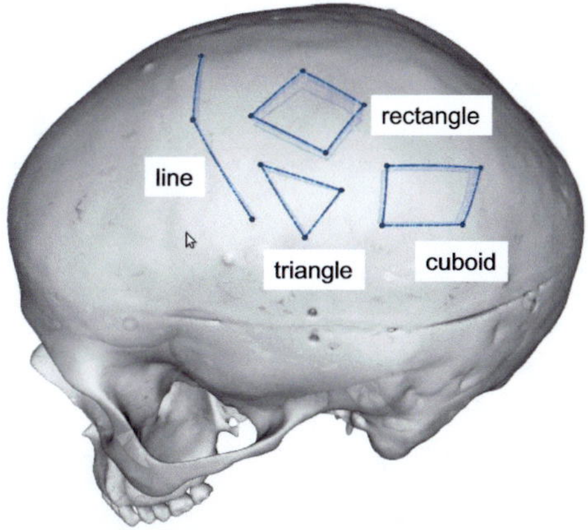

Figure 7.1: Predefined marking paths on the surface model of a human skull replica.

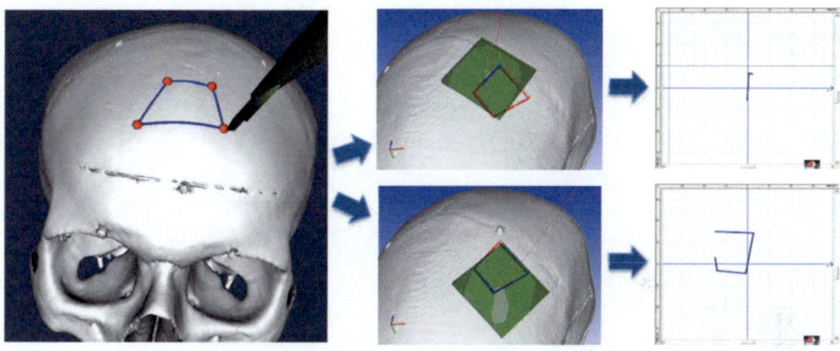

Figure 7.2: The geometrically planned marking path (left) is transferred into two scan head locations (middle) and the corresponding ablation patterns (right).

of 0.46 mm for the line, 0.47 mm for the triangle and 0.42 mm for the cuboid.

After registration the robot positions the scan head into the first location relative to the skull. After processing of the according marking job with the laser the robot moves into the next location and processing of the next marking job part is performed. The remaining part is then

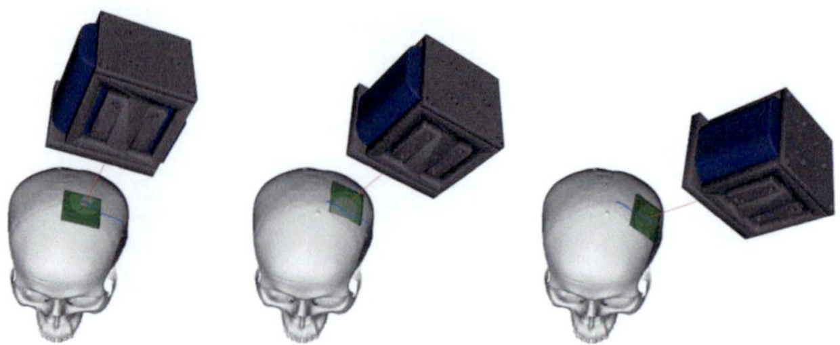

Figure 7.3: Three scan head locations are necessary in order to process the line marking path.

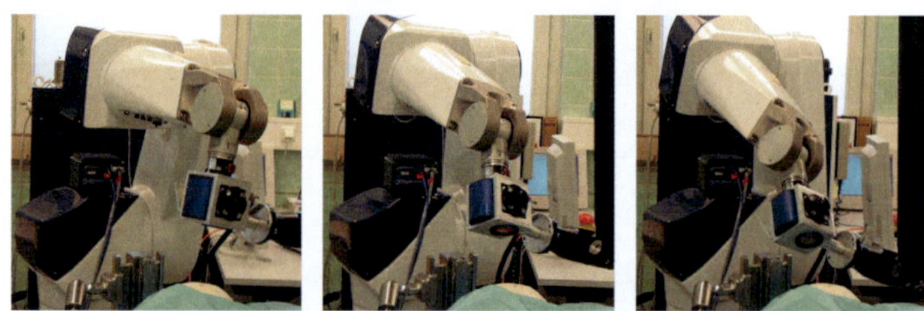

Figure 7.4: For processing the predefined line marking path the robot positions the scan head relative to the skull in the three optimized locations. Compare Figure 7.3.

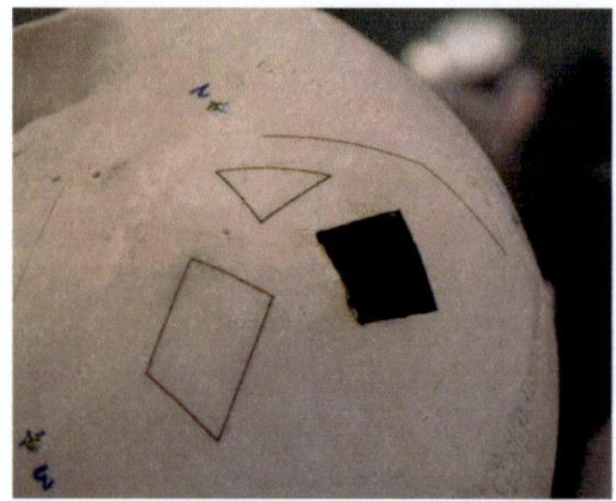

Figure 7.5: Resulting markings on the skull. The craniotomy is not further evaluated since the hard plastic showed material changes after melting due to increasing heat by laser irradiation.

executed in the last scan head location. Figure 7.4 illustrates the robot poses during the execution of the line marking path.

The laser marking was in all three cases performed with a pulse repetition of 400 Hz and a pulse duration of 28 µs. Figure 7.5 illustrates the skull after processing of all experiments. Beside the three marking paths a craniotomy is visible in the picture. Since the hard plastic melts under longer laser irradiation due to the increasing heat, this craniotomy is not evaluated. The distorted edges and rehardening material would falsify the results.

7.1.4 Evaluation

For evaluation the trajectories on the skull were verified with a FARO measurement arm. Using a point probe (accuracy of 0.005 mm) the fiducial locations were measured first. Afterwards the support points of the marking paths were measured. In the cases of the triangle and the cuboid the measured points correspond to the vertices. For the line the support point resulting from the planning could be measured, since they indicate a change in the path direction. The measurement data was registered to the CT dataset of the skull utilizing the fiducials and a point based registration method. All measured points were transformed into the planning coordinate system for further evaluation. The distances between the planned support point location and the measured support point were determined and the deviations were calculated. Furthermore the marking path lengths between two succeeding support points were compared to the corresponding planned and measured points.

7.1.5 Results

Registration between the FARO measurement and the planning coordinate system results in a FRE of 0.288 mm. Table 7.1 summarizes the deviations between planned and measured support points and segment lengths for all three cases. Furthermore the mean deviation in the support point location and the segment lengths is given.

The mean positioning error is 0.49 mm and the mean deviation for length is 0.175 mm. The root mean square error (RMSE) for positioning is 0.306 mm and for lengths 0.082 mm. Higher deviations (max. error) than 0.71 mm were not recognized in these experiments.

Table 7.1: Deviations between planned and measured support points and segment lengths for all three cases (all values in mm).

Support	Cuboid		Triangle		Line	
Point	ΔPoint	ΔLength	ΔPoint	ΔLength	ΔPoint	ΔLength
1	0.612	0.028	0.425	0.127	0.719	0.107
2	0.774	0.258	0.299	0.124	0.495	0.155
3	0.471	0.228	0.549	0.172	0.169	-
4	0.388	0.377	-	-	-	-
Mean	0.561	0.223	0.424	0.141	0.461	0.131

7.2 Cutting Experiments

In order to evaluate the overall system capabilities cutting experiments were performed in the scope of this doctoral thesis. To reflect the parameters of human bone two types of animal bone were chosen: femoral cow bone and lower jaw bone of a pig. Both specimen are characterized by their bone compactness. Furthermore the thickness of both bony specimen is comparable to human bones.

7.2.1 Femoral Cow Bone

In the following the experiment performed on a femoral cow bone preparation will be explained[1].

Preparation

A fresh ex-vivo femoral cow bone was prepared using a band saw with a diamond coated blade. The resulting block measures about 60 mm × 50 mm × 40 mm. Four titanium marker screws for registration were added before acquiring a surface scan using a FARO Laser ScanArm. The measured point cloud was triangulated (cp. Figure 7.6a) and the resulting surface model was used for the further planning procedure.

Planning

A quadric block was defined by specifying four support points on the surface which correspond to the corners (cp. Figure 7.6b) as well as corresponding points at the underside of the bone. The planned cutting

---

[1]This experiment was published in [Bur09c].

trajectory has a maximum depth of 5.356 mm, a minimum depth of 4.591 mm and a median depth of 5.098 mm.

After transferring the geometrical definition into a corresponding ablation pattern an optimized amount of scan head locations was determined automatically. In this case two robot locations are necessary in order to process the laser cutting (cp. Figure 7.6c and d). The second scan head location accounts for the thickness of the bone and therefore relocates the focal point position inside the cut.

Execution

For the cutting experiment the femoral cow bone specimen was fixated using a Mayfield clamp like device (cp. Figure 7.6e. The gold standard method for registration purposes was applied: Determining the

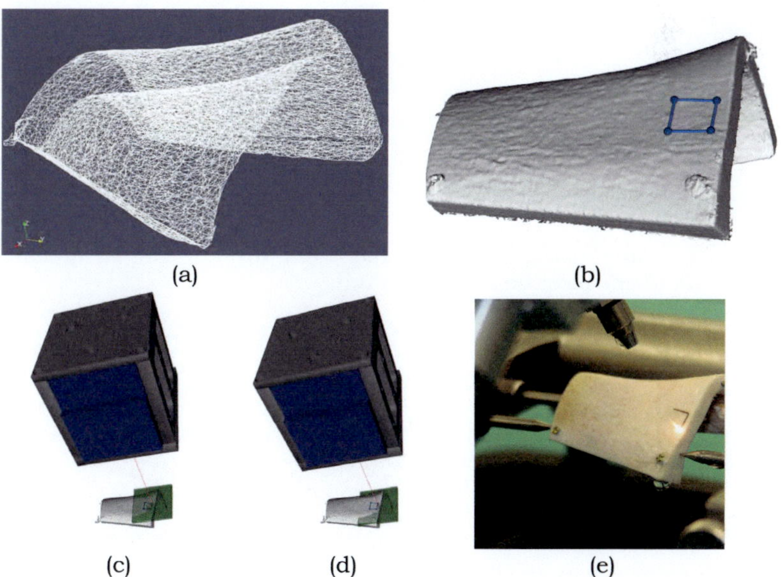

Figure 7.6: (a) Triangulated surface from the laser scan. (b) Planned cutting trajectory. (c) First robot location for execution. (d) Second robot location for execution. (e) Processing of the trajectory.

titanium marker screw locations in coordinates of the surface model (planning coordinate system) and acquiring the screw locations in the experimental setup using a measurement arm (Microscribe G2X, Immersion Inc., San Jose, USA). The registration had an FRE of 0.54 mm.

Overall the cutting experiment was performed successfully. After executing the preplanned ablation patterns in the corresponding scan head locations the osteotomy was not possible, since the bone was not cut through. This is due to the fact that depth development of laser ablation (cp. Section 5.2.2) was not considered for this initial experiment. Performing the same ablation pattern in the second scan head location once again, leads to the final cut through. The quadric bone piece could be removed, as shown in Figure 7.7. The cutting width at the surface is about 400 µm.

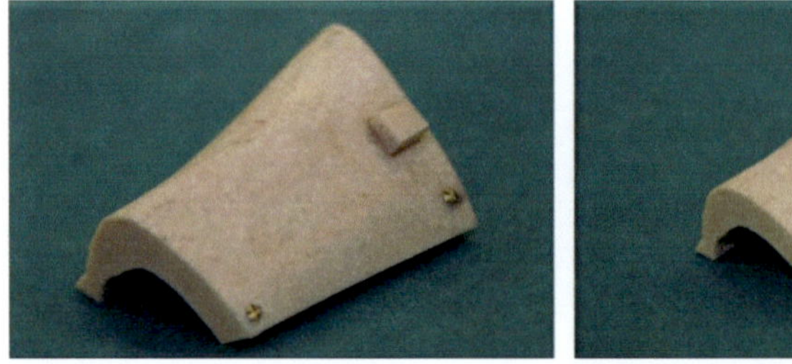

Figure 7.7: After performing the experiment the preplanned quadric bone piece could be removed.

7.2.2 Cadaver Skull

The second experimental cutting series was performed on a lower jaw of a pig and is explained in the following.

Preparation

A fresh ex-vivo half-skull of pig was prepared by removing the scalp and isolating the lower jaw. Four titanium marker screws (referred to as fiducials in the following) were implanted (cp. Figure 7.9a), before acquiring a CT dataset at the Department for Radiology of the Municipal Clinic in Karlsruhe, Germany. The CT was acquired with a slice distance of 0.6 mm and a resolution in x-/y-direction of 0.293 mm. The bone was segmented in the CT datasets automatically and the location of the fiducials were determined.

Planning

Three cutting trajectories were planned for the experimental cutting series. Figure 7.8 illustrates the planned cuts. In the following the cuts will be referred to as triangle, cuboid and rectangle according to their shape.

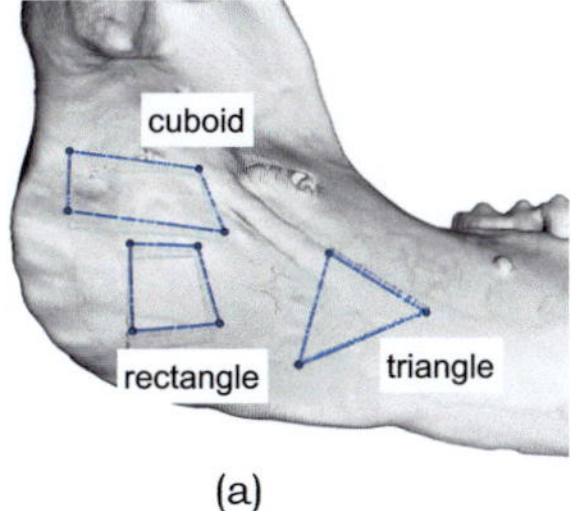

(a)

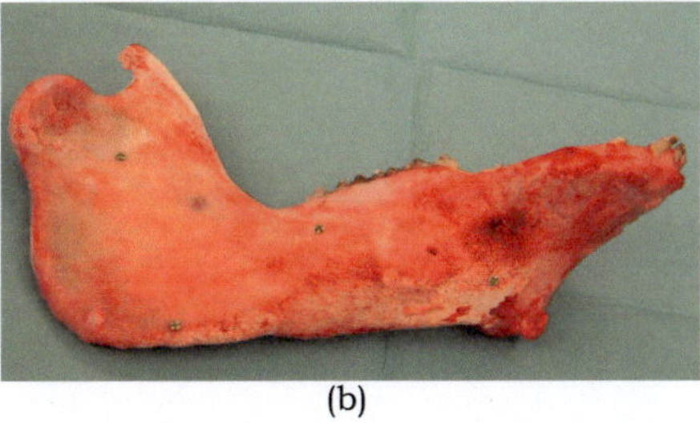

(b)

Figure 7.8: (a) Three cutting trajectories were preplanned for the experimental cutting series. (b) Isolated lower jaw of a fresh ex-vivo pig equipped with four titanium marker screws for registration.

The cutting geometries were automatically transferred into their corresponding ablation patterns. Based on these, the optimization algorithm stated in Section 5.3 was utilized in order to determine the minimum amount of scan head locations for processing. The rectangle requires two scan head locations, while the triangle has to be processed with five and the cuboid two scan head repositionings. The amount of repositionings corresponds to the cutting depth.

Experiment

For the experiment the jaw bone was fixated on an operating table. Figure 7.9(b) illustrates the experimental setup. The location of the fiducials was measured with a FARO measurement arm. The registration between the CT coordinate system and the intraoperative coordinate system was determined. After performing one cutting trajectory, the bone was relocated on the operating table in order to achieve three independent intraoperative situations (i.e. three different registrations).

The resulting fiducial registration errors (FRE) are stated in Table 7.2, as well as parameters of the cuts. Figure 7.10 illustrates the experimental results.

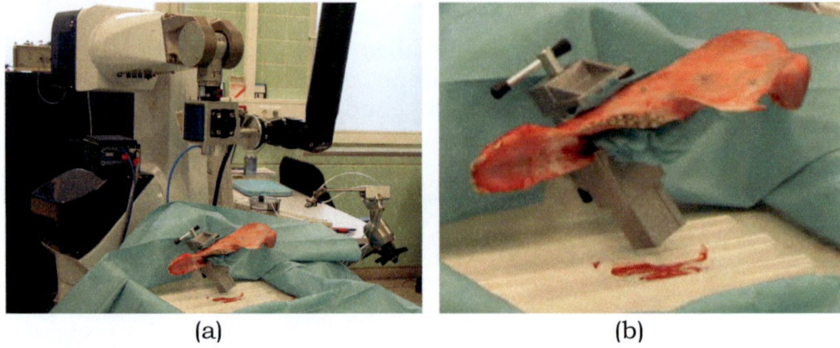

(a)　　　　　　　　　　　　　　　(b)

Figure 7.9: (a) Experimental setup. (b) Detail enlargement.

Table 7.2: Planned cutting trajectories parameters.

	Cuboid	Triangle	Rectangle
Maximum depth [mm]	8.870	19.8686	7.335
Minimum depth [mm]	2.739	7.824	5.371
Median depth [mm]	5.388	13.411	3.553
Trajectory length [mm]	80.612	69.993	65.531
No. of repositionings	2	4	1
FRE [mm]	0.523	0.264	0.403
Pulse repetition rate [Hz]	400	400	400
Pulse duration [µs]	200	160	160
Line speed [mm/sec]	40	80	80
Processing time [min]	53	120	150

7.2.3 Evaluation

For the evaluation the lower jaw was frozen after finishing the experiments and defrosted shortly before acquiring a postoperative CT dataset at the Department of Radiology of the Municipal Clinic in Karlsruhe, Germany. The CT was acquired with a slice distance of 0.6 mm and a resolution in x-/y-direction of 0.293 mm. The bone was automatically

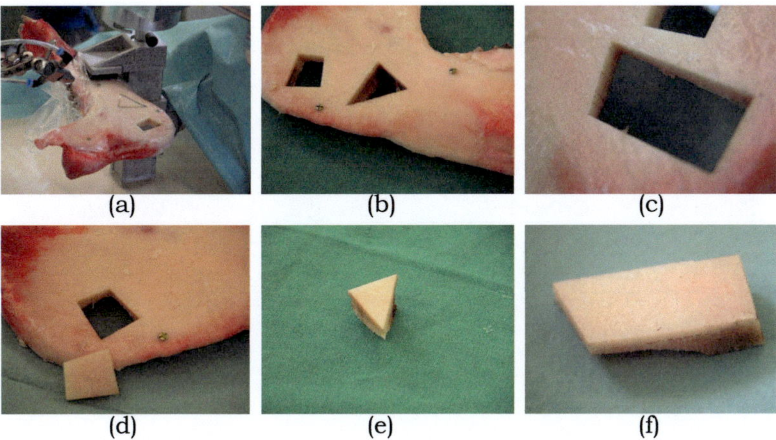

Figure 7.10: (a) Processing of the triangle. (b) Triangle cut. (c) Cuboid cut. (d) Rectangle cut. (e) Bone piece removed after successful triangle cut. (f) Bone piece removed after successful cuboid cut.

segmented and the fiducial locations were determined manually. The CT dataset was transformed into the preoperative planning coordinate system by applying point-based registration between the pre- and postoperative fiducial locations. The FRE of the registration was 0.167 mm.

In the transformed postoperative CT dataset the locations of the support points of the three cuts were determined. The distances between the planned and the measured support point locations were determined and the deviation was calculated. Furthermore the marking path length between two succeeding support points was compared between the planned and measured points.

7.2.4 Results

The deviations between planned and postoperative measured support points and segment lengths for all three cutting trajectories performed on the isolated lower jaw are stated in Table 7.3. Furthermore the mean error in positioning and segment lengths is stated for each experimental cut. Errors greater than 0.76 mm were not recognized. The overall mean positioning error is 0.392 mm and the mean deviation for length is 0.179 mm. The root mean square error (RMSE) for the position of the cut is 0.154 mm and for the length deviation 0.411 mm.

Table 7.3: Deviations between planned and measured support points and segment lengths for all three cuttings performed at the isolated lower jaw (all values in mm).

Support Point	Cuboid		Triangle		Rectangle	
	ΔPoint	ΔLength	ΔPoint	ΔLength	ΔPoint	ΔLength
1	0.468	0.100	0.567	0.015	0.448	0.189
2	0.401	0.145	0.548	0.760	0.271	0.093
3	0.351	0.099	0.277	0.178	0.263	0.177
4	0.502	0.022	-	-	0.213	0.194
Mean	0.431	0.092	0.464	0.318	0.299	0.163

7.3 Medical Application

Beside the evaluation of the developed methods for robot assisted laser osteotomy in general, laser based cochleostomy as a specific medical application was considered in the scope of this doctoral thesis. In collaboration with the Department of Otorhinolaryngology, Head and Neck Surgery of the University Hospital Düsseldorf, Germany a method to support accurate cochlear implantation (cp. Figure 7.11) was developed. Cochlear implantation is the method of choice for giving deaf people, especially children, back the acoustic sensation. Several researchers are currently aiming at supporting the surgical procedure of cochlear implantation by means of computer and robot assistance (see also Section 3.2.5). However, none of the proposed methods found its way into clinical routine yet. Therefore, further considerations have to be undertaken.

In the scope of this doctoral thesis the developed methods for robot assisted laser osteotomy were adapted for robot assisted cochlear implantation. Experimental trials for laser based cochleostomy were performed[2].

7.3.1 Cochleostomy

The cochleostomy is one step during cochlear implantation. A small bore hole is drilled into the cochlear for insertion of the implant's electrode. Precise bone ablation is a key issue for the minimal traumatic cochleostomy. Especially to preserve residual hearing the protection of the cochlear's lining membrane (endost) is mandatory. In a laboratory setup it was already shown, that it is possible to preserve boundary layers during laser ablation using visual control on the basis of image processing techniques [Kah08, Kah09a]. In order to evaluate operating room feasibility the developed methods for accurate positioning of the laser by means of robot assisted surgery were conducted in an experiment.

7.3.2 Preparation

An ex-vivo human temporal bone from the Department for Anatomy of the University Hospital Düsseldorf was used for the setup trial. The

---

[2]Planning and simulation of laser based cochleostomy was published in [Kah09b]. An experimental setup trial was published in [Bur09b].

temporal bone specimen was prepared conventionally with a mastoidectomy and posterior tympanotomy by an ENT surgeon. Four titanium screws were inserted as fiducials for registration. Afterwards a CT dataset was acquired with a slice distance of 2 mm and resolution in x-/y-direction of 0.238 mm.

The temporal bone was segmented automatically in the CT dataset. Utilizing the marching cubes algorithm, a three dimensional surface model was generated for planning. Furthermore the locations of the fiducials were determined manually. Figure 7.12 illustrates the CT images of the temporal bone specimen.

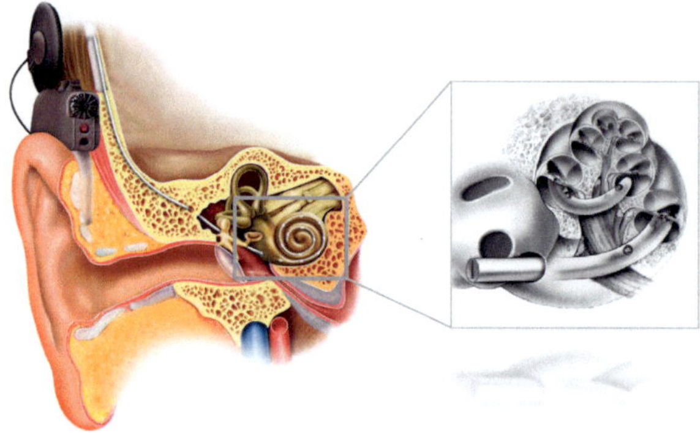

Figure 7.11: Illustration of a cochlear implant [MED09a]. The electrode is inserted into the cochlear through a small bore hole. The opening of the cochlear could be achieved by laser bone ablation.

7.3.3 Planning

After defining the target point on the promontory of the cochlear and the desired angle (respectively the end point in the scala tympani), the diameter of the cochleostomy canal was defined. The cochleostomy volume is about 1-2 mm^3. Figure 7.13 illustrates planning and simulation of robot assisted laser based cochleostomy.

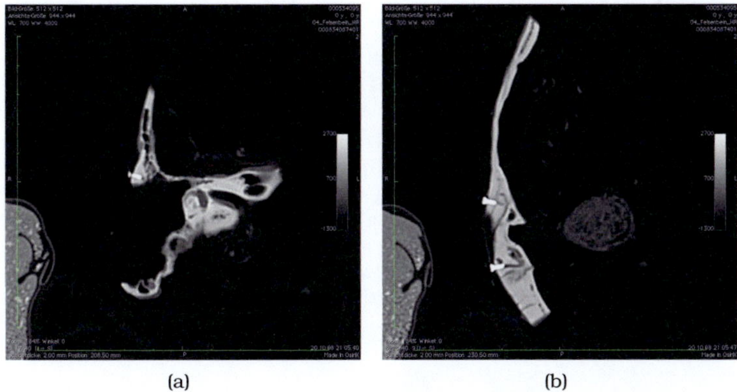

(a)   (b)

Figure 7.12: CT images of (a) human temporal bone with mastoidectomy and posterior tympanotomy and (b) fiducial markers.

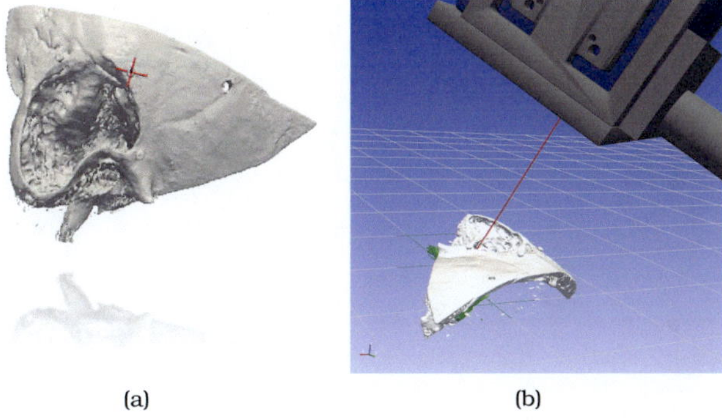

(a)   (b)

Figure 7.13: (a) The cochleostomy is planned on a three-dimensional surface model of a human temporal bone specimen prepared conventionally with a mastoidectomy and posterior tympanotomy. The cross illustrates a segmented titanium screw for registration. (b) After defining the target point on the promontory of the cochlear and the desired angle the robot location is planned and simulated.

7.3.4 Setup Trial

For the experimental setup trial the bone specimen was fixated on the operating table in a Mayfield clamp like device (cp. Figure 7.14b). The human temporal bone specimen was registered, by measuring the four fiducial markers with the Microscribe measurement arm. Applying a point-based registration algorithm in order to determine the spatial correspondence between the planning coordinate system and the experimental setup specific situation results in a FRE (fiducial registration error) of 0.37 mm. Applying the registration information to the preplanned cochleostomy results in the relative location to the temporal bone in robot coordinates. Figure 7.14a illustrates the setup trial. The pilot laser beam augments the target point onto the promontory.

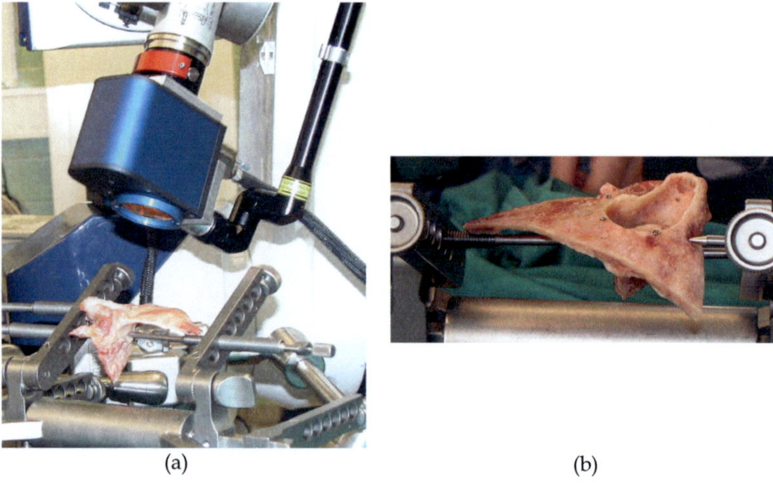

(a)   (b)

Figure 7.14: (a) The robot moves into the preplanned location after applying the registration information. (b) The temporal bone specimen was fixated for the experiment in a Mayfield clamp like device.

7.4 Conclusion

The experimental feasibility studies revealed the applicability of the developed workflow for robot assisted laser osteotomy. Marking experiments on a human skull replica could be performed successfully with an overall positioning accuracy of 0.49 mm. The cutting experiments result in a mean overall positioning accuracy of less than 0.4 mm. Considering the error which develops over the complete workflow this value is exciting. Starting with the CT image data acquisition which already brings inaccuracy into the overall system by limited resolution in the slice distance (0.6 mm) and the succeeding patient model determination which also includes errors during the segmentation and triangulation step, planning is performed on this somehow inaccurate model. Another source of inaccuracy is the registration step: The manual error in segmentation of the fiducial location in the CT images and the measuring error in localizing the fiducials shortly before the experiment come together in this workflow part.

FRE errors ranging from 0.25 mm to 0.55 mm already indicate the resulting positioning error of the cut. It is important to notice, that the mean deviation of the marking and cutting lengths with 0.17 mm and 0.19 mm indicates that the main error contributes to the absolute positioning of the cut but has less influence on the relative positioning. The repeatability accuracy of the used robot with 0.02 mm and the scan head with 20 µrad are assumed to have negligible impact on the overall accuracy.

As stated in Section 4.3.3 the variance in the ablated depth per pulse varies between -3 % and +7 % from the mean ablation depth. Therefore the preplanned ablation pattern can only be regarded as a prediction for the ablation process. The performed experiments revealed that more or less pulses may be necessary in order to achieve the breaktrough for osteotomy. An appropriate online sensing technique for controlling the ablation process, either in energy per pulse or number of pulses, is indispensable for robot assisted laser osteotomy, if stuctures that have to be protected are present behind the bone.

The experimental setup trial is an important step towards robot assisted laser based cochleostomy feasible for the OR. For the first time laser ablation and robot assistance are combined in order to support the surgeon during the microsurgical and most crucial part in cochlear implantation. Using laser ablation allows less traumatization due to a smaller bore channel towards the tympanic cavity due to smaller handling volume compared to conventional burs. The combination of the presented setup in this doctoral thesis with our established video surveillance technique for visual control of the ablation process will allow for protection of the cochlear's lining membrane.

The experimental evaluation of the robot assisted laser osteotomy system and workflow developed in the scope of this thesis verified the feasibility and applicability of the methods. Even without any online sensing technique during execution the accuracy achieved is excellent. However, each step of the workflow allows further optimization and therefore the achieved accuracy can be regarded as an upper boundary for robot assisted laser osteotomy. The experimental results proved the high potential of computer and robot assisted laser ablation for highly accurate surgical applications.

8

Discussion and Outlook

Every discovered energy source had a great, if not massive impact on our way of processing material. So in surgery: While humans already opened the skull with means of sharped flints ten thousand of years before Christ, surgical tools developed from stones fastened to wooden handles over bronze blades to mechanical instruments as they are used in clinical practice nowadays. The invention of laser radiation in the 1960s and the possibility to apply highly focused energy which induces changes in matter drove the development of applications and specialized laser systems for material processing, both in industry and in medicine. For example, lasers in opthalmic surgery or dermatology fundamentally promoted medical care and opened up new medical procedures. Using a *laser scalpel* in order to cut soft tissue in surgery is more the less state of the art today.

Right after the development of the first laser systems, researchers tried to cut bony tissue using laser. Suitable wavelengths in the infrared which result in strong absorption in the mineral or water component of the bone were identified. However, the chosen laser parameters indicate that cutting bone leads to thermal injury (combined with carbonization) and is therefore not feasible. Around the 1990s it could be shown, that short laser pulses are suitable for effective hard tissue ablation. Recently the feasibility of short-pulsed CO_2 laser ablation for bone processing was proven in histological examinations. Hence, a new cutting method is available which promises a new era in bone treating medical applications. Contact- and force-free processing of complex incisions with two tenth of a millimeter in width are now possible to achieve. However, this precision exceeds manual practicability.

Therefore the main objective of this work was to develop a robotic system which enables precise bone cutting by means of laser ablation. The main contribution of this doctoral thesis is the establishment of world's first system for robot assisted laser osteotomy. Methods of computer and robot assisted surgery were reconsidered and composed to a workflow for laser osteotomy, which was completely implemented in the scope of this doctoral thesis. Adequate calibration and registration methods are proposed, which allow precise execution. Hence, for the first time robot assisted processing of a predefined cutting incision using short-pulsed CO_2 laser ablation with an overall accuracy below 0.4 mm can be achieved. Thus, the scientific key question (cp. Section 3.4) *How to realize laser cutting of bone with robotic methods?* is answered.

The state of the art revealed an insufficiently described laser ablation process. In the scope of this doctoral thesis the development of a laser induced incision was investigated based on the concatenation of single laser pulses. For the first time, laser induced incision profiles were evaluated in cutting direction utilizing a confocal microscope. It was shown, that laser ablation does not allow to gain straight cutting edges offhand. Furthermore, several laser process parameters were determined, which are crucial for modeling the ablation process in order to answer the posed scientific question: *Which parameters are essential for controlling and optimizing the ablation process?* In particular the inclination angle was analyzed in regard to its impact on the ablation crater. Application of a two-dimensional scan head for deflecting the laser beam onto the beam was chosen and process parameters were determined. This is mainly the distance of the bone to the focusing lens. It could be proven, that within the Rayleigh Distance around the focal point position effective laser ablation is achievable.

Based on the findings regarding the laser ablation process and cut development by pulse concatenation, adequate planning methods were achieved in the scope of this doctoral thesis. This allows predefinition of cuts on a patient specific model. Geometrical definitions of incisions can be automatically transferred into a sequence of single laser pulses now. Furthermore methods for optimized procedure planning are contributed with this doctoral thesis. The optimal amount of scan head locations is automatically determined by solving the minimum set cover problem using a modified greedy algorithm. For each scan head locations the corresponding two-dimensional coordinates for the single laser pulses are generated. This allows automatic execution of robot assisted laser osteotomy for the first time. Hence, answers were found for the scientific question: *How to plan hard tissue ablation preoperatively?*

In order to answer the question *Which cutting accuracies are required and obtainable with a system for robot assisted laser osteotomy?*, experimental feasibility studies proved the applicability of the developed workflow for robot assisted laser osteotomy. Cutting experiments on ex-vivo animal bone result in a mean overall accuracy of less than 0.4 mm. Considering the error which develops over the complete workflow this value is exciting. Steps towards robot assisted laser based cochleostomy feasible for the OR could be made in an experimental setup trial. For the first time laser ablation and robotic assistance are combined in order to support the surgeon during the microsurgical and most crucial part in cochlear implantation.

8.1 Discussion

In order to highlight the main contributions of this doctoral thesis, the results and achievements are discussed from a medical and a technical point of view.

At a first glance industrial laser applications, where a robot is utilized in order to position the laser, seem to already provide the solution of how to realize laser bone cutting. However, the methods of robot assisted material processing using laser are not adaptable offhand to medical applications. Motivated by remote laser welding applications, where a robot positions a scan head relatively to the workpiece, a system was realized for robot assisted laser osteotomy utilizing a two-dimensional scan head. The scan head deflects the laser beam fast and precisely in the working area. A robot could not achieve such high accuracy. In addition, such fast robot movements would not be suitable for applications in the operation theater, where the patient and the medical staff is in proximity to the robot. But a two-dimensional scan head necessitates exact knowledge of the working area as well as the characteristics of the scan plane which is directly correlated to the applied focus lens. As described in Section 6.3, the calibration and registration of the involved coordinate systems and assurance of laser beam alignment are essential system parameters, deciding whether precise laser bone ablation in micrometer range is achievable or not. The methods contributed by this doctoral thesis are feasible and experiments revealed high accuracies.

Optimized procedure planning for laser ablation is a contribution of this doctoral thesis as well. For the first time cutting of predefined complex trajectories using laser ablation is realized. Planning on a patient specific three-dimensional model derived from image acquisition allows to preoperatively define the incision. Thanks to the developed process model an approximation of the amount of laser pulses to be applied in order to achieve the defined cutting depth is now possible. However, the ablation depth is dependent on the individual bone composition as well as the assistive fluid applied. Therefore, planning can just provide an estimation for the online process. This is especially useful, when applying online sensing techniques for controlling the process of bone ablation. The basis for that is contributed by the achievements of this doctoral thesis. Additionally a simulation environment allows to visualize the treatment in advance. Especially collision free execution and reachability are important issues, since the applied passive articulated mirror arm for beam delivery is implicitly moved by the robot.

From the medical point of view, the realized robotic system allows for the first time application of short-pulsed CO_2 laser ablation to cut bony tissue in the operation theater. The contact-free cutting method is preferable, since the bone is not exposed to mechanical forces, affected by metal abrasion etc. Furthermore the induced thermal injury is negligible and laser induced incisions are known to heal faster than these achieved with conventional instruments.

Complex and arbitrary cutting geometries are now feasible. This opens up absolutely new possibilities for processing bony tissue. The reduced bone loss and thereby considerably smaller cutting width allow for ostoetomy geometries which are self-stabilizing. For example, cutting out puzzle like geometries and their relocation for osteosynthesis without necessarily stabilizing the bone with plates is now a conceivable scenario. In comparison conventional cutting instruments are known with smallest diameter of 0.6 mm. Usually diameters according to the width and diameter of burrs and saws of around 1 mm are applied in clinical practice depending on the application.

Furthermore the ablation by single laser pulses, each removing tiny bone fragments in micrometer range facilitates precise microsurgical treatment of bone. Accuracies unachievable by manual laser application, either with or without support of a navigation system, are obtainable with the proposed robotic approach. Considering the state of the art this was inconceivable up to now.

However, the drawback of short-pulsed CO_2 laser ablation compared to conventional cutting instruments is the low processing speed. Until now no method was introduced which increases the processing speed not involving broadening the incision. Further developments regarding the laser technology and adjustment of laser parameters inherit the potential for optimizing the processing speed. For example, a related diploma thesis promises that the repetition rate can be quintupled when adapting the distance between successive laser pulses according to thermal considerations [Zha09]. In-vivo studies and histological analysis have to prove this assumption. Additionally application of catalyzing fluids other than water is promising as introduced in Section 4.4.4.

The cutting experiments carried out in the scope of this doctoral thesis showed a relatively large median incision depths. The number of pulses necessary for increasing incision depths grows exponentially (cp. Figure 5.4). Nevertheless, the advantages of cutting bony tissue using laser have to be weighed against the processing speed for each medical application. As an example: An experienced surgeon needs approximately half an hour to perform a craniotomy of standard dimension, but without the possibility to execute it in a preplanned way. Actually, laser processing would take longer depending on the bone thickness, but would guarantee all mentioned advantages (e.g. less bone loss, precise execution, exact realization of a predefined cut). Furthermore the slower processing could be balanced by a less time expensive osteosynthesis procedure, since application of plates for fixation may not be necessary anymore.

Nowadays, there are four principal means of cutting bony tissue: the conventional mechanical cutting, the water jet method, piezoelectric cutting and finally laser cutting. While the advantages and disadvantages of mechanical cutting were already discussed in the scope of this thesis, water jet and piezoelectric cutting are two competing cutting methods against laser cutting and therefore shortly described in the following in order to complete the view on bone cutting methods:

Water jet cutting is established since the 1980s in some medical fields, such as visceral surgery [Kuh05]. The method utilizes water jet in addition with biocompatible abrasives for effective bone cutting. The principal clinical feasibility could be shown for water jet cutting. The advantages over conventional mechanical cutting, such as non contact process and free cutting geometries, face the disadvantages of decreasing quality with increasing cutting depth and problematic control of cutting depth. Furthermore the parameters of water jet cutting have to be carefully chosen. While water jet cutting could not establish as a clinical method for bone cutting, it is used for cutting soft tissue, such as nerves and vessels.

Piezoelectric cutting utilizes micro-vibrations of scalpels at ultrasonic frequencies for dissecting bony tissue. Also known as Piezosurgery®, the technology was invented by Vercellotti in 2004 [Ver04] and is produced by Mectron (Carasco, Italy). Low modulated ultrasonic frequency is used to induce micro–vibrations of the tool between 60 and 210 µm. Hence, selective cutting can be achieved, since only specific tissue responds to the ultrasonic frequency, such as hard tissue at 25-29 kHz. There is no explicit force needed for cutting. However, if too high forces are used, the vibration of the tool is impeded and leads to development of heat. Histologically less traumatization could be proven, but there are also remarks on poor histological sharpness with irregular cut lines [Rom09]. Selective cutting makes this method favorable especially to preserve critical structures [Egg04, Sal08], but it is slow and not reasonable for deep cuts.

Table 8.1 compares the four presented bone cutting methods to each other. Each bone cutting method shows specific advantages compared with the others. Hence, one cannot implicate one favorable method. In fact, the method has to be carefully chosen in respect of the medical application. Regarding osteotomy, laser cutting is in an inferior position to piezoelectric cutting from an economic point of view. How-

Table 8.1: Comparison of existing bone cutting methods:
(++) excellent, (+) good, (-) minor, (--) poor.

	Mechanical	Water Jet	Piezoelectric	Laser
Width	> 0.6 mm	~1 mm	⌀tool + 0.2 mm	µm range
Geometry	limited	free	tool dependent	free
Selective	--	-	++	+
Trauma	--	+	+	+
Force	--	++	+	++
Degree of Popularity	++	--	-	+
Pro	clinical practice	integrated cooling	preserves critical structures	microsurgical bone removal
Contra	abrasion	no control of cutting depth	slow	slow

ever, laser bone cutting has the strong advantage of microsurgical bone removal, which is not achievable with the other methods. Recent histological comparison studies also indicate the highest bone cut precision for laser cutting [Rom09].

8.2 Outlook

In the scope of this doctoral thesis answers to the scientific key questions posed in Section 3.4 are provided. However, finding one answer in science raises new questions. In the following an outlook is given regarding further research issues and medical application of robot assisted laser bone ablation.

The contributions of this doctoral thesis raise two further main scientific questions, which are essential to investigate before applying robot assisted laser osteotomy in a real medical application. This is on the one hand concerning cutting strategies and on the other hand concepts for online measurement of the ablation process.

To overcome the drawback of comparably slow processing speeds, it is essential to particularly review cutting strategies, the kind of laser system and laser parameters. It could be shown that analyzing a laser induced incision based on the concatenation of single laser pulses is of high importance in order to understand the development of a cut. Additionally the variation and adjustment of laser parameters as well as the composition of the bone directly influence the amount of bone removed by a single laser pulse. On this basis cutting strategies can be developed. Further research has to be focused on the analysis of cutting profiles according to the placement of laser pulses under consideration of the effective thermal dispersion. In the scope of this thesis the laser parameters proposed in the state of the art were used and changed. It is therefore essential to proof biological compatibility histologically. Here, experimental evaluation with in-vivo animal and human bone are obligatory.

In this context further research regarding the applied laser is also important. Since the developed methods in the scope of this doctoral thesis are adaptable to other laser systems which ablate bony tissue thermo-mechanically, application of further laser types is feasible. Regarding the findings of Section 4.5.3 an interesting aspect for laser processing of bony tissue lies in the variance of the beam profile. For example application of specific beam shaping lens systems to achieve a quadratic beam profile could lead to more straight cutting edges (e.g. [Hom09]). However, future application of short-pulsed CO_2 laser for bone cutting could utilize Q-switched laser systems in order to reduce high variances in the output power and therefore variances in the ablated volume per pulse.

Beside the investigation of cutting strategies which could enhance the processing speed, further research regarding assistive fluids seems promising (cp. Section 4.4.4). Application of fluids other than water in order to catalyze the ablation process has not been in the scope of research up to now. Pretreatment of the bone in order to weaken the composite structure can lead to increased volumes removed by single laser pulses.

The second scientific challenge to deal with in future is the online surveillance of the laser cutting procedure. This is essential especially for those applications which require preservation of underlying vital structures (e.g. blood vessels, nerves) or bone lining membranes (e.g. dura mater, endost). In the state of the art video camera based control of laser ablation for preserving the inner lining membrane of the cochlear during cochleostomy were already proposed [Kah09a] and could be applied to the setup realized in the scope of this doctoral thesis. However, more accurate and reliable detection of remaining bony tissue to cut could be achieved by applying optical coherence tomography (OCT) as online sensing technique. Based on the measurement the laser process parameters can be adjusted. Both online sensing methods could be directly coupled into the existing optical path, i.e. coaxial coupling using a beam combiner before the deflecting mirrors in the two-dimensional scan head.

As a consequence of applied online measurement methods, the process model proposed in this doctoral thesis could be adapted online, according to the observed ablation performance. It is imaginable that before starting the cutting itself, a few laser pulses are applied onto the bone and directly measured in their dimension. Attributing this patient specific information to the ablation model would lead to a more reliable estimation of the process itself (adaption of the plan) and therefore imply conditions for the online process control. In order to enhance the explication of the ablation process model including stochastic parameters which account for bone composition parameter and process variances is another option.

Beside these two most important scientific challenges further optimization of the system itself can be performed. In a first step a three-dimensional scan head should be applied. This would allow for adjusting the focal point of the beam for every laser pulse and thereby lead to more constant ablation results and cause less repositioning by the robot. Main objective towards applicability in the operation theater should be the integration of the laser beam guidance into the robot kinematics itself (cp. Votan C-BIM introduced in Section 2.4.1). Here, it is also imaginable not to apply a scan-head anymore, but to integrate the focus lens directly to the robot's last joint and to integrate a beam combiner for direct coupling of measurement devices. In this context both, the cutting strategy and risk analysis have to decide whether the movement of the robot in the required speed is suitable under OR conditions or not.

Short term goal is the application of the realized system for robot assisted laser osteotomy in the operation theater and the execution of a surgical intervention. As long as no online measurement method is introduced to the system, it is possible to evaluate the system capabilities by performing marking paths with the laser intraoperatively. This step towards an intraoperative application is essential in order to present this new methodology to surgeons and to discuss further optimization needs, especially regarding the compatibility with the surgical workflow.

Further discussions with surgeons of different disciplines will reveal new medical application areas. For example in cranio-maxillofacial surgery: the fronto-orbital advancement (FOA) which requires osteotomy of multiple bone fragments and their realignment is a promising application field for laser bone cutting. Accurate execution of predefined cuts and the minimized bone loss are significant advantages. In addition the bone thickness is comparably small since this operation is accomplished on young children. Another promising medical application area is orthopedics. The precise bone removal could be advantageous for preparing individual implant beds.

To conclude, laser bone ablation is going to gain significance in micro-surgical applications treating bony and cartilage tissue. Furthermore it is anticipated, that laser bone ablation will revolutionize bone cutting, if the processing speed is optimized. The basis for robot assisted application of laser for osteotomy is laid with this doctoral thesis.

Risk Analysis

In the scope of this doctoral thesis a complete risk analysis of the realized system for robot assisted laser osteotomy (cp. Chapter 6) was performed in order to support the development process. The main purpose was to detect hazards of the system in an early development stage and be able to reduce or even eliminate risks by appropriate measures. With regard to commercialization, the risk analysis was performed according to the harmonized European standard ISO 14971 and its annexes [DIN07]. Here, only the Fault Tree Analysis (FTA) is presented. In an associated diploma thesis the comprehensive risk analysis is given, including the Failure Mode and Effect Analysis (FMEA) [Wim09]. In the following the three steps of the risk analysis are described.

A.1 Step 1: Intended Use and Identification of Medical Device Characteristics that Could Impact Safety

A.1.1 Intended Use

The robot assisted laser osteotomy system realized in the scope of this thesis is intended to be clinically used for ablation of human bone. The main characteristics of the system are:

- Preoperative computer assisted planning of cutting trajectories and simulation of the intervention

- Automatic positioning of a two-dimensional beam deflector (scan head), which is attached to the flange of a robot's end-effector, by a robot

- Bone ablation using the laser beam as surgical tool by deflecting it onto the target area using a scan head in order to perform a preplanned cutting trajectory.

A.1.2 Reasonably Foreseeable Misuse

All applications that differ from the intended use are considered to be reasonably foreseeable misuses, including:

- Moving of the robot while laser is turned on

- Processing of other material than bone (e.g. metal or plastic)

- Using under other than specified environmental conditions (e.g. in an explosive atmosphere).

A.1.3 Identification of Safety-Related Characteristics

The consideration of criteria which refer to the safety of a medical device are important for the identification of hazards associated with the medical device. Following the questions of Appendix C in [DIN07] the intended use, intended users, foreseeable misuse and the functions of the individual components of the system and their interactions in normal as well as fault state need to be considered.

A.2 Step 2: Identification of Hazards

The best source of information on hazards associated with a device is to look for similar devices and their hazards. Online complaint databases for medical devices offer a large amount of publications reporting malfunctions of medical devices and injury of patients and users by medical devices. Here the Medical Product Safety Network (MedSun) [Med09b], which is a reporting programme launched by the U.S. Food and Drug Administration (FDA), the Laser Accident Database, maintained by Rockwell Laser Industries [Roc09], USA, and the German Federal Institute for Drugs and Medical Devices (*Bundesinstitut für Arzneimittel und Medizinprodukte BfArM*) [BfA09] were regarded to get ideas about hazards associated with the components of the experimental setup.

The Fault Tree Analysis (FTA) of the risk analysis deals with the identification of hazards using fault trees. The four main hazards to the system are graphically analyzed using fault event and gate shapes to illustrate, top-down, the process that might lead to the main failure.

Dangers associated with the use of lasers can generally be categorized into primary and secondary dangers. The primary hazard is the laser beam itself, which can affect humans and objects. The laser beam can thereby be raw, focused, directly reflected or scattered. Therefore primary hazards are laser-related hazards.

Secondary hazards can be direct or indirect (following [Bar06]):

- Direct secondary dangers are caused by components of the complete laser system (device related),

- Indirect hazards are generated by interaction of the radiation with materials or the atmosphere (application related).

For the developed system for robot assisted laser osteotomy the following main risks were identified and decomposed using FTA:

- Collision of robot with patient,

- Injury of patient caused by laser radiation,

- Injury of patient caused by secondary laser hazards (device or application related),

- and Wrong incision.

In order to make the fault trees clearly represented, three sub events are defined using transfer gates to link them to the main hazards:

- Wrong, invalid or inaccurate registration,

- Unexpected movement of patient,

- and Unintended acceptance of robot motion.

Figures A.1 to A.7 show the corresponding fault trees for robot assisted laser osteotomy.

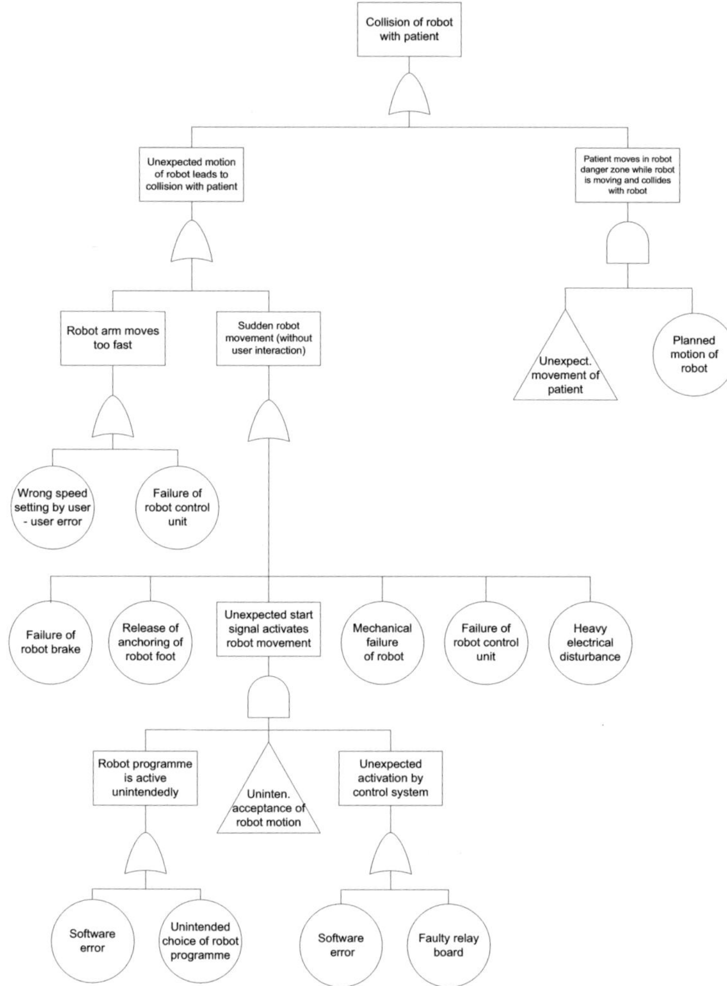

Figure A.1: Collision of robot with patient.

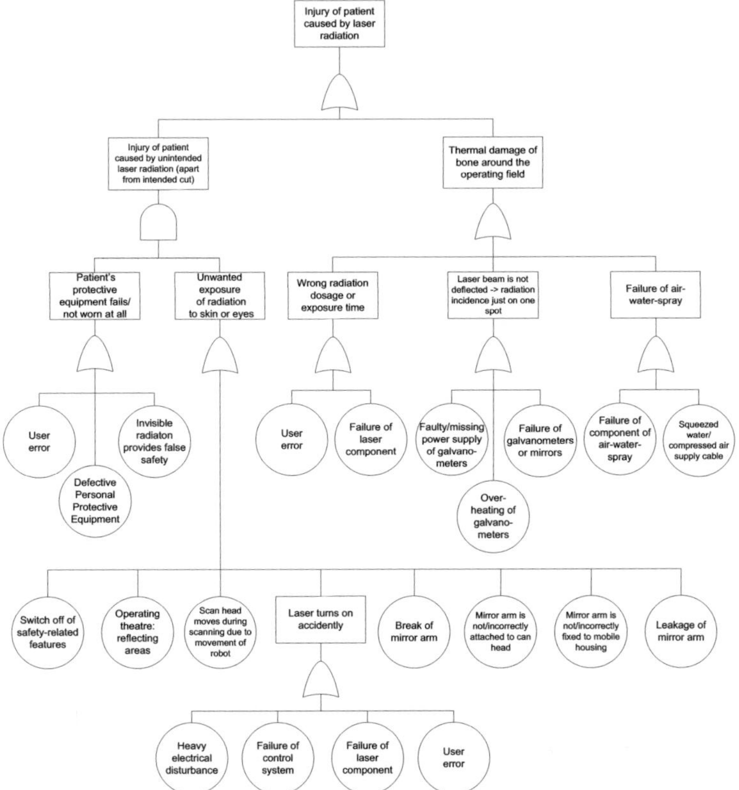

Figure A.2: Injury of patient caused by laser radiation.

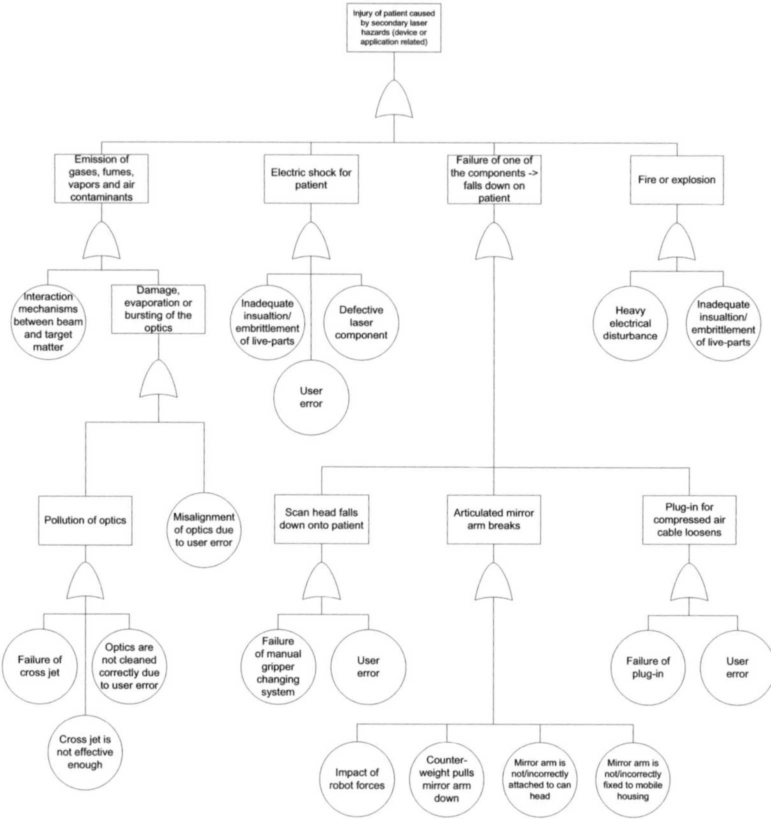

Figure A.3: Injury of patient caused by indirect laser hazards.

Figure A.4: Wrong incision.

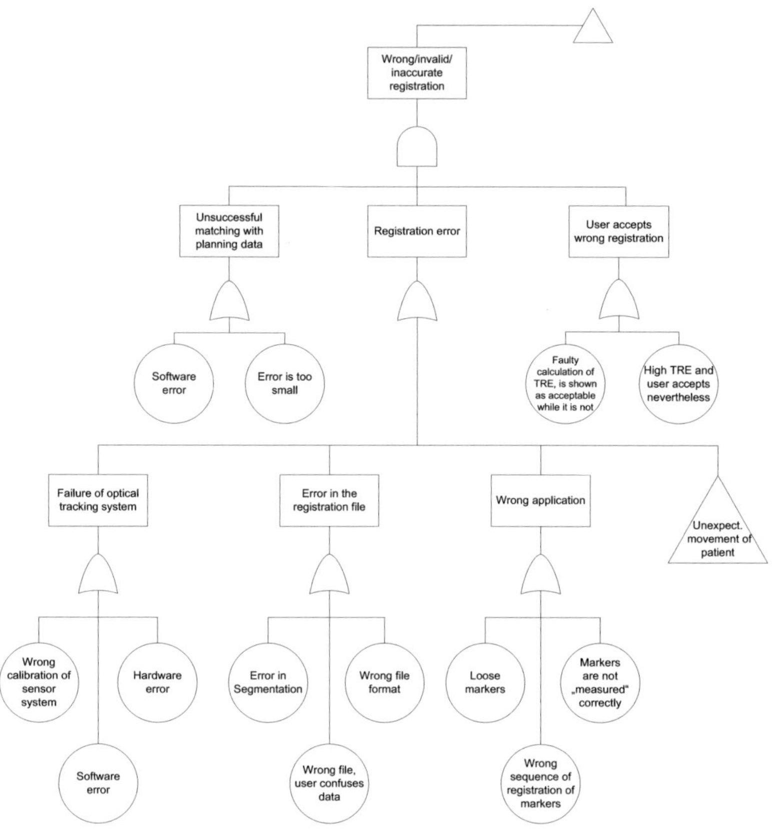

Figure A.5: Wrong/invalid/inaccurate registration.

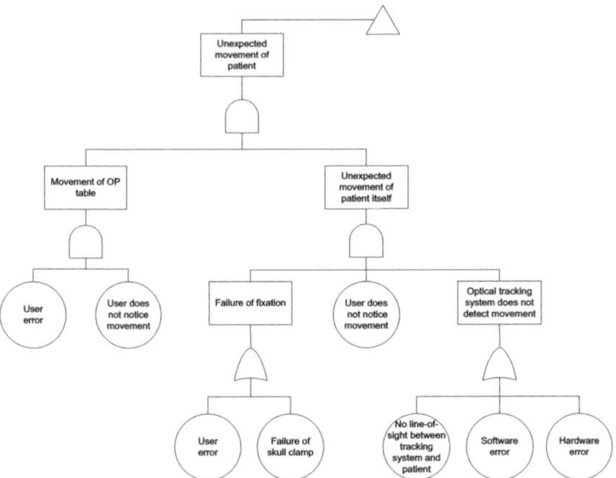

Figure A.6: Unexpected movement of patient.

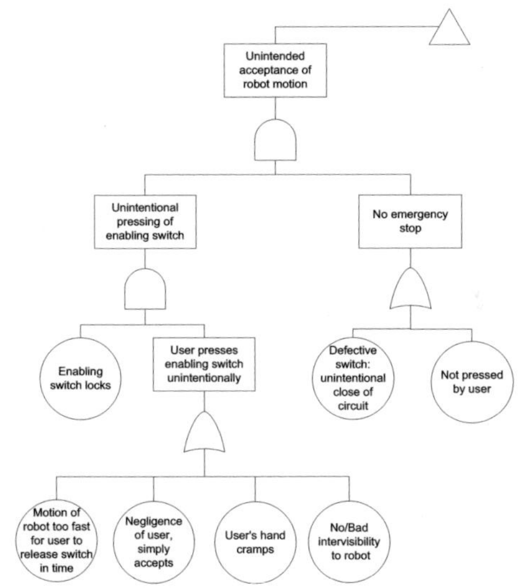

Figure A.7: Unintended acceptance of robot motion.

A.3 Step 3: Risk Evaluation and Risk Control

After analyzing the potential risks of the system for robot assisted laser osteotomy, risk evaluation and risk control are performed. Therefore a risk evaluation for every identified cause of a hazard based upon a comparison between the estimated risks and predefined risk acceptability levels is assessed first. In the next step measures for risk mitigation are proposed.

A.3.1 Risk Evaluation

The risk of each identified cause of a hazard is attributed to one of the three defined risk acceptability levels in the risk graph, i.e. *broadly acceptable*, *ALARP* (as low as reasonable possible) and *intolerable*. Overall 200 risks were identified. Figure A.8 depticts the overall risk graph.

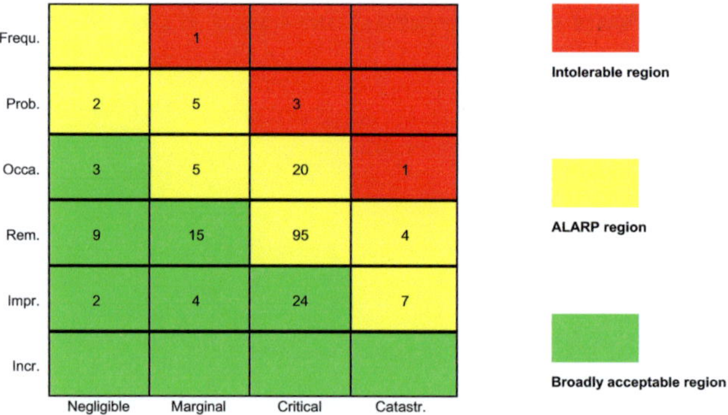

Figure A.8: All risks of the system for robot assisted laser osteotomy assigned to the risk graph according to their severity.

Five causes of hazards are ranked to the red, intolerable region. Two of them are user-related causes and may happen due to the invisible radiation of the used CO_2 laser, which results in a false sense of security. First, the risk of unwanted exposure of radiation to the skin or the eyes is considered as critical and probable. Second, the fact that personal protective equipment (e.g. laser radiation protective goggles) is not used or is inappropriate for the laser in use is also evaluated as critical and probable. Third, a fire in the OR due to the ignition of inflammable objects or material by the laser beam is considered as catastrophic and occasional. Fourth, a wrong laser incision resulting from wrong or missing calibration of the robot is considered as critical and probable. Last but not least, the severity of a harm occurring due to a position change of the robot joint angles, is estimated as marginal, but supposed to happen frequently because of the robot design.

The mentioned unacceptable risks necessitate reduction to the ALARP or broadly acceptable region by mitigating the severity of harm and the probability of occurrence. If this is not possible, the decision of application has to be based on a risk-benefit analysis , whether the medical or surgical procedure is worth to take the risk for patient compared with the possible benefit, if the procedure is successful.

137 hazards are assigned to the ALARP region. They need to be further investigated and, if possible, reduced. The 58 risks assigned to the broadly acceptable region can be accepted without further analyses.

To conclude, the most profound hazards associated with the realized setup for robot assisted laser osteotomy are related to the use of the CO_2 laser in the operating theater. Especially user errors can affect the adequate use of the laser and of the other system components. This may endanger the user himself, the patient, medical staff and the environment. Since user-related errors can never be entirely eliminated, the main priority is to reduce them to a minimum by trying to make the system inherently safe and by implementing appropriate safety measures. Furthermore, critical hazards are associated with component failures. The use of a robot in the OR constitutes a further threat. Collisions with humans as well as unexpected movements of the robot during the laser ablation process are considered critical, too.

A.3.2 Risk Control

In order to reduce risks associated with the intolerable and ALARP region in the risk graph, measures can be taken. Risk control measures can generally be divided into three categories: inherently safe design, safeguards or protective measures and warnings. In the following some examples for measures of each category are presented:

Inherently Safe Design

The most profound hazards associated with the experimental setup for robot assisted laser osteotomy are related to the use of the CO_2 laser in the operating theater. The main safety criterion is therefore to shield the laser system to operate it hazard free and to protect operators and patients from harmful unintended radiation of the CO_2 laser. The manufacturer has already incorporated safety precautions into the laser to make it intrinsically safe. Furthermore, in order to prevent the articulated mirror arm from falling down on the patient, it is hold in an upward position by utilizing a spring balancer under the ceiling (cp. 6.2.2). This spring balancer is attached to the ceiling by a carabiner and additionally secured by a steel wire rope.

Safeguards or Protective Measures

An important safeguard to be installed is an emission exhaust system in order to capture by-products of the laser processing and utilization of a room ventilation system, that should be available in every OR. Further proposed safeguards include flow controllers for monitoring the compressed air and the cooling water of the air-water spray for process optimization. In order to alert the OR personnel, an activation warning system (i.e. a warning light or an audible alarm during laser processing) should be installed.

Warnings

Warnings are considered as weakest measures, as they only inform the operator about residual risks, leaving the compliance to the user. Nevertheless, communicating about hazards or improper use, for example, by providing adequate education and training to the OR personnel, is important in order to alert the user of potential hazards, e.g. about the invisible CO_2 laser radiation. Thus, the suggested administrative control measures include warning signs and labels, that designate laser and robot use areas.

B

Mirror Arm Specification

Various design possibilities allow to apply an optimized articulated arm for a certain application. The quality of an articulated arm is strongly dependent on its calibration. However, the quality usually declines over the life cycle, since the straight tubes are affected by bending. Furthermore the overall weight of the arm as well as the stiffness of the joints have an impact on the overall accuracy.

For realization of the system for robot assisted laser osteotomy in the scope of this doctoral thesis, an industrial articulated mirror arm was specified in strong collaboration with the manufacturer Laser Mechanisms. Figure B.1 and B.2 show the CAD drawing of the applied mirror arm.

Six mirror joints and an additional rotational last joint were chosen in order to allow positioning with 7 degrees of freedom. With an overall weight of $\approx 20\,\mathrm{kg}$, a length of $2231\,\mathrm{mm}$, a tubes diameter of $70\,\mathrm{mm}$ and stiff joints the arm provides a high accuracy, which is of high importance for the application. A spring balancer is utilized in order to keep the arm in an upward position and reduce the affecting weight at the robot's end-effector.

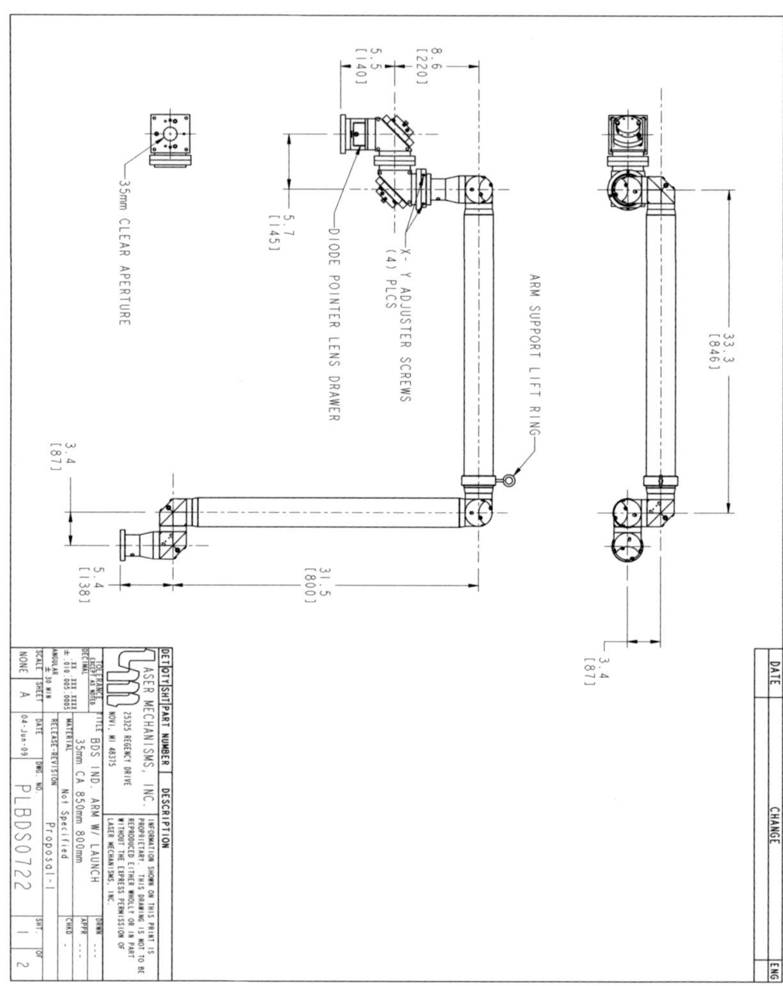

Figure B.1: CAD drawing of the mirror arm. Courtesy of Laser Mechanisms Inc. [Las09]

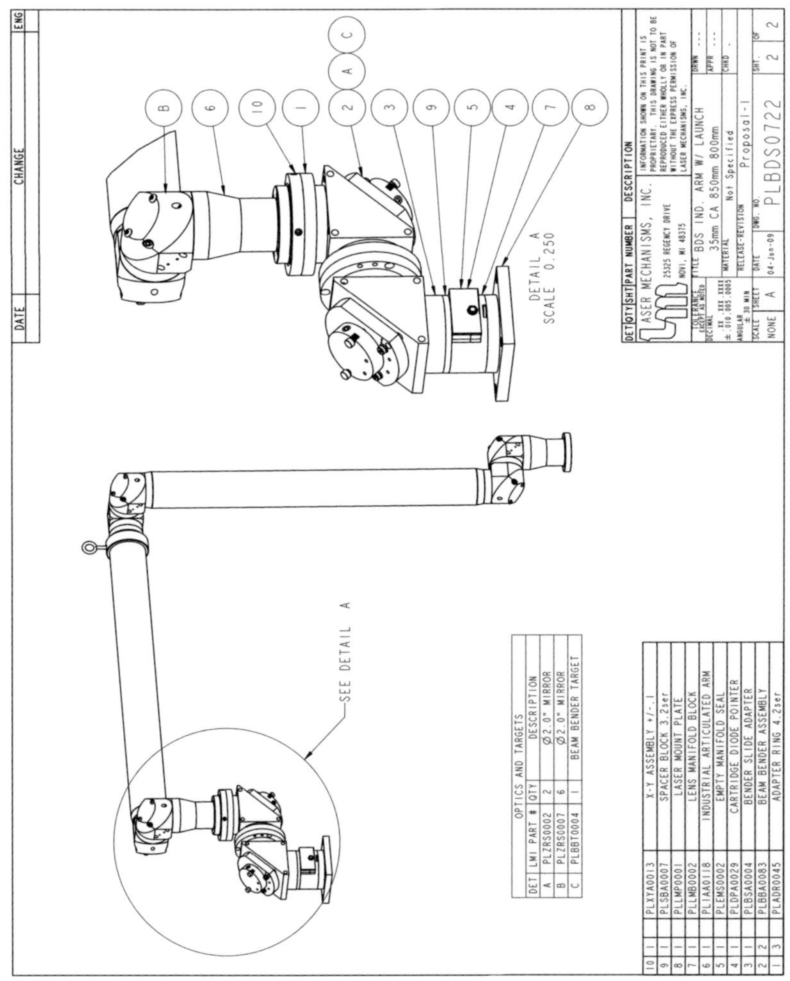

Figure B.2: CAD drawing of the mirror arm. Courtesy of Laser Mechanisms Inc. [Las09]

C

Robot Specifications

C.1 Stäubli RX90B CR

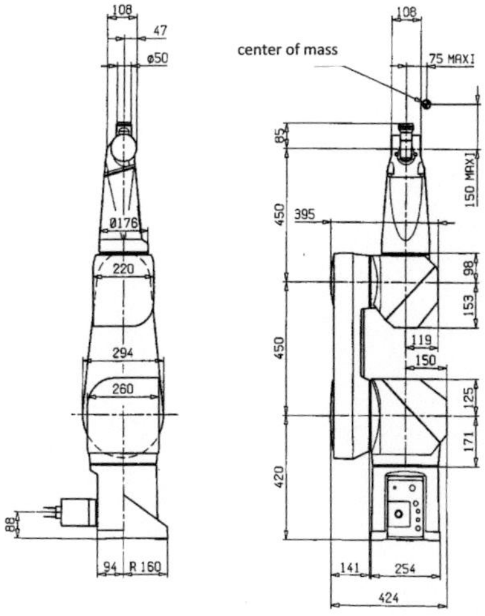

Figure C.1: Dimensions of the Stäubli RX90B robot. All values in mm. [Stä04]

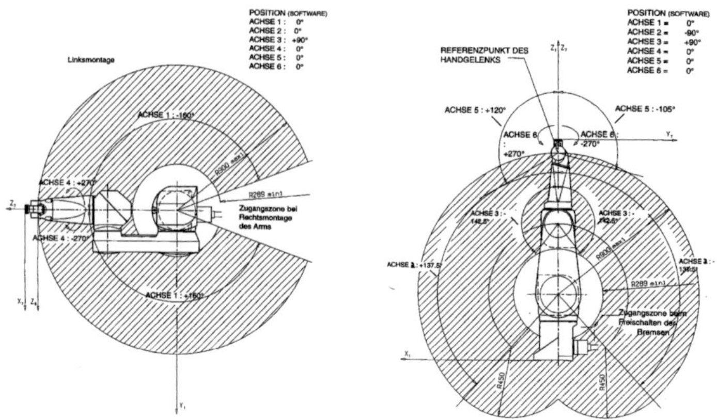

Figure C.2: Workspace of the Stäubli RX90B robot. All values in mm.
[Stä04]

C.2 KUKA LWR

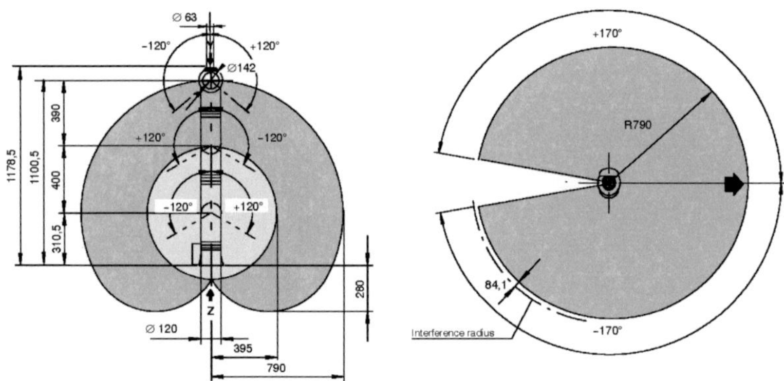

Figure C.3: Workspace and dimensions of the KUKA LWR robot. All
values in mm. [KUK07]

Bibliography

[Als06]   Y. Alsultanny. "Laser Beam Analysis Using Image Process-
            ing". *Journal of Computer Science*, Vol. 2, No. 1, pp. 109–113,
            2006.

[Ama02]   E. Amara and A. Bendib. "Modelling of vapour flow in deep
            penetration laser welding". *Journal of Physics D: Applied
            Physics*, Vol. 35, No. 3, pp. 272–280, 2002.

[And01]   H. J. Andersen. *Sensor Based Robotic Laser Welding*. Doc-
            toral thesis, Aalborg University, Denmark, 2001.

[Ari07]   A. Arif and B. Yilbas. "Thermal stress developed during the
            laser cutting process: consideration of different materials".
            *The International Journal of Advanced Manufacturing Tech-
            nology*, Vol. 37, No. 7, pp. 698–704, 2007.

[Arn71]   J. Arnaud, W. Hubbard, G. Mandeville, B. de la Clavière,
            E. Franke, and J. Franke. "Technique for Fast Measure-
            ment of Gaussian Laser Beam Parameters". *Applied Optics*,
            Vol. 10, No. 12, pp. 2775–2776, 1971.

[Bad06]   C. Bader and I. Krejci. "Indications and limitations of Er:YAG
            laser applications in dentistry.". *American Journal of Den-
            tistry*, Vol. 19, No. 3, pp. 178–186, 2006.

[Ban08]   I. N. Bankman. *Handbook of Medical Image Processing and
            Analysis*. Academic Press, 2008.

[Bar06]   K. Barat. *Laser safety management*. Vol. 107 of *Optical sci-
            ence and engineering*, CRC Taylor & Francis, 2006.

[Bas06]   P. Bast, A. Popovic, T. Wu, S. Heger, M. Engelhardt,
            W. Lauer, K. Radermacher, and K. Schmieder. "Robot- and
            computer-assisted craniotomy: resection planning, implant
            modelling and robot safety.". *The International Journal of
            Medical Robotics and Computer Assisted Surgery*, Vol. 2,
            No. 2, pp. 168–178, 2006.

[Bäu04]   D. Bäuerle, H. Bergmann, F. Dausinger, K. Dörschel, A. Gebhardt, M. Geiger, M. Grupp, H. Haferkamp, C. Hertzler, H. Hügel, O. Minet, M. Möhrle, G. Müller, W. O'Neill, W. Schulz, G. Sepold, H. Tiziani, M. Totzeck, M. Ulbricht, H. Venghaus, F. Vollertsen, H. Welling, and W. Wiesemann. *Landolt-Börnstein: Numerical Data and Functional Relationships in Science and Technology - New Series; Group 8: Advanced Materials and Technologies: Laser Physics and Applications*. Vol. 1, Subvolume C, Springer, 2004.

[Bei03]   L. Beiser. *Unified Optical Scanning Technology*. Wiley-Interscience, 2003.

[Ben87]   A. Benabid, P. Cinquin, S. Lavalle, J. Le Bas, J. Demongeot, and J. de Rougemont. "Computer-Driven Robot for Stereotactic Surgery Connected to CT Scan and Magnetic Resonance Imaging". *Applied Neurophysiology*, Vol. 50, No. 1-6, pp. 153–154, 1987.

[Ber03]   H. Berlien and G. Müller. *Applied laser medicine*. Springer, 2003.

[Beu00]   J. Beutel, H. Kundel, and R. V. Metter. *Handbook of medical imaging: Physics and Phychophysics*. Vol. 1, SPIE Press, 2000.

[Bey03]   E. Beyer, A. Klotzbach, V. Fleischer, and L. Morgenthal. "Nd:YAG - Remote Welding with Robots". In: *Proceedings of the WLT-Conference on Lasers in Manufacturing*, pp. 367–373, 2003.

[Bey98]   E. Beyer and K. Wissenbach. *Oberflächenbehandlung mit Laserstrahlung. Laser in Technik und Forschung*, Springer, 1998.

[BfA09]   BfArM Bundesinstitut für Arzneimittel und Medizinprodukte. online: `http://www.bfarm.de`, Apr. 2009.

[Bog08]   R. Bogue. "Cutting robots: a review of technologies and applications". *Industrial Robot: An International Journal*, Vol. 35, No. 5, pp. 390 – 396, 2008.

[Bol05]   G. Bolmsjo and M. Olsson. "Sensors in robotic arc welding to support small series production". *Industrial Robot: An International Journal*, Vol. 32, No. 4, pp. 341–345, 2005.

[Bor07]   R. Boris, D. Eun, A. Bhandari, K. Lyall, M. Bhandari, C. Rogers, O. Alassi, and M. Menon. "Potassium-titanyl-phosphate laser assisted robotic partial nephrectomy in a porcine model: can robotic assistance optimize the power needed for effective cutting and hemostasis?". *Journal of Robotic Surgery*, Vol. 1, No. 3, pp. 185–189, 2007.

[Bou86]   J. Boulnois. "Photophysical processes in recent medical laser developments: a review.". *Laser in Medical Science*, Vol. 1, pp. 47–66, 1986.

[Bre07]   P. N. Brett, R. P. Taylor, D. Proops, C. Coulson, A. Reid, and M. V. Griffiths. "A surgical robot for cochleostomy". In: *Proceedings of the IEEE Engineering in Medicine and Biology (EMB)*, pp. 1229–1232, 2007.

[Bru02]   L. Bruzzone, R. Molfino, and R. Razzoli. "Modelling and Design of a Parallel Robot for Laser-cutting Applications". In: *Proceedings of the IASTED International Conference Modelling, Identification and Control (MIC)*, pp. 518–522, 2002.

[Bur08]   J. Burgner, J. Raczkowsky, and H. Wörn. "Establishment of an Experimental Setup for Robot Assisted Laser Osteotomy with Corresponding Simulation Environment for Optimization of all Relevant Parameters". In: *Proceedings of the Annual Meeting of the German Society for Computer and Robot Assisted Surgery (CURAC)*, pp. 147–148, 2008.

[Bur09a] J. Burgner, L. A. Kahrs, J. Raczkowsky, and H. Wörn. "Including Parameterization of the Discrete Ablation Process into a Planning and Simulation Environment for Robot-Assisted Laser Osteotomy". In: *Medicine Meets Virtual Reality 17 - NextMed: Design for/the Well Being*, Vol. 142 of *Studies in Health Technology and Informatics*, pp. 43–48, IOS Press, 2009.

[Bur09b] J. Burgner, F. Knapp, L. A. Kahrs, J. Raczkowsky, H. Wörn, J. Schipper, and T. Klenzner. "Setup and experimental trial for robot-assisted laser cochleostomy". *International Journal of Computer Assisted Radiology and Surgery*, Vol. 4 (Suppl. 1), pp. 118–119, 2009.

[Bur09c] J. Burgner, M. Müller, J. Raczkowsky, and H. Wörn. "Robot assisted laser bone processing: Marking and cutting experiments". In: *Proceedings of the International Conference on Advanced Robotics (ICAR)*, pp. 1–6, 2009.

[Bur09d] J. Burgner, J. Raczkowsky, and H. Wörn. "End-effector calibration and registration procedure for robot assisted laser material processing: Tailored to the particular needs of short pulsed CO_2 laser bone ablation". In: *Proceedings of the IEEE International Conference on Robotics and Automation (ICRA)*, pp. 3091–3096, 2009.

[Cas00] C. Casadei, S. Martelli, and P. Fiorini. "A Workcell for the Development of Robot-Assisted Surgical Procedures". *Journal of Intelligent and Robotic Systems*, Vol. 28, No. 4, pp. 301–324, 2000.

[Che03] A. K. Cherri and M. S. Alam. "Accurate measurement of small Gaussian laser beam diameters using various rulings". *Optics Communications*, Vol. 223, No. 4-6, pp. 255–262, 2003.

[Che07] M.-F. Chen and Y.-P. Chen. "Compensating technique of field-distorting error for the CO_2 laser galvanometric scanning drilling machines". *International Journal of Machine Tools and Manufacture*, Vol. 47, No. 7-8, pp. 1114–1124, 2007.

[Cho00] S. Choi and W. Newman. "Design and Evaluation of a Laser-Cutting Robot for Laminated, Solid Freeform Fabrication". In: *Proceedings of the IEEE International Conference on Robotics and Automation (ICRA)*, pp. 1551–1556, 2000.

[Cla78] L. Clayman, T. Fuller, and H. Beckman. "Healing of continuous-wave and rapid superpulsed, carbon dioxide, laser-induced bone defects". *Journal of Oral and Maxillofacial Surgery*, Vol. 36, No. 12, pp. 932–937, 1978.

[Cle01] K. Cleary and C. Nguyen. "State of the art in surgical robotics: clinical applications and technology challenges.". *Computer Aided Surgery*, Vol. 6, No. 6, pp. 312–328, 2001.

[Cle05a] K. Cleary. "Medical Robotics and the Operating Room of the Future". In: *Proceedings of the IEEE Engineering in Medicine and Biology (EMB)*, pp. 7250–7253, 2005.

[Cle05b] K. Cleary and A. Kinsella. "OR 2020: the operating room of the future.". *Journal of Laparoendoscopic & Advanced Surgical Techniques*, Vol. 15, No. 5, pp. 495–500, 2005.

[Cle06] K. Cleary, A. Melzer, V. Watson, G. Kronreif, and D. Stoianovici. "Interventional robotic systems: applications and technology state-of-the-art.". *Minimally Invasive Therapy and Allied Technologies*, Vol. 15, No. 2, pp. 101–113, 2006.

[Cob06] C. M. Cobb. "Lasers in Periodontics: A Review of the Literature". *Journal of Periodontology*, Vol. 77, pp. 545–564, 2006.

[Cor01] T. Cormen. *Introduction to algorithms.* MIT Press, 2001.

[Cou08] C. Coulson, R. Taylor, A. Reid, M. Griffiths, D. Proops, and P. Brett. "An autonomous surgical robot for drilling a cochleostomy: preliminary porcine trial". *Clinical Otolaryngology*, Vol. 33, No. 4, pp. 343–347, 2008.

[Cra89] J. Craig. *Introduction to robotics: mechanics and control.* Addison-Wesley, 2nd Ed., 1989.

[Cru07] R. A. C. Cruces and J. Wahrburg. "Improving robot arm control for safe and robust haptic cooperation in orthopaedic procedures". *The International Journal of Medical Robotics and Computer Assisted Surgery*, Vol. 3, No. 4, pp. 316–322, 2007.

[Dah08] N. Dahotre and S. Harimkar. *Laser Fabrication and Machining of Materials.* Springer, 2008.

[Dav06] B. Davies, M. Jakopec, S. J. Harris, F. Rodriguez y Baena, A. Barrett, A. Evangelidis, P. Gomes, J. Henckel, and J. Cobb. "Active-Constraint Robotics for Surgery". *Proceedings of the IEEE*, Vol. 94, No. 9, pp. 1696–1704, 2006.

[Dav91] B. L. Davies, R. D. Hibberd, W. S. Ng, A. G. Timoney, and J. E. A. Wickham. "A surgeon robot for prostatectomies". In: *Proceedings of the International Conference on Advanced Robotics (ICAR)*, pp. 871–875, 1991.

[deM08] E. de Mello, R. Pagnoncelli, E. Munin, M. Filho, G. de Mello, E. Arisawa, and M. de Oliveira. "Comparative histological analysis of bone healing of standardized bone defects performed with the Er:YAG laser and steel burs". *Lasers in Medical Science*, Vol. 23, No. 3, pp. 253–260, 2008.

[Del06]   H. Delingette, X. Pennec, L. Soler, J. Marescaux, and N. Ayache. "Computational Models for Image-Guided Robot-Assisted and Simulated Medical Interventions". *Proceedings of the IEEE*, Vol. 94, No. 9, pp. 1678–1688, 2006.

[Des08]   S. Desai, C.-K. Sung, D. Jang, and E. Genden. "Transoral Robotic Surgery Using a Carbon Dioxide Flexible Laser for Tumors of the Upper Aerodigestive Tract". *The Laryngoscope*, Vol. 118, No. 12, pp. 2187–2189, 2008.

[Deu83]   T. F. Deutsch and M. W. Geis. "Self-developing UV photoresist using excimer laser exposure". *Journal of Applied Physics*, Vol. 54, No. 12, pp. 7201–7204, 1983.

[DIN07]   DIN EN ISO 14971:2007. "Medical devices – Application of risk management to medical devices". 2007.

[DIN08]   DIN EN ISO 11145. "Optics and photonics – Lasers and laser-related equipment – Vocabulary and symbols". 2008.

[Dix06]   G. Dixin, C. Jimin, and C. Yuhong. "Laser Cutting Parameters Optimization Based on Artificial Neural Network". In: *Proceedings of the IEEE International Joint Conference on Neural Networks*, pp. 1106–1111, 2006.

[Du04]   D. Du, B. Sui, Y. F. He, Q. Chen, and H. Zhang. "Visual Measurement of Path Accuracy for Laser Welding Robots". In: *Robotic Welding, Intelligence and Automation*, Vol. 299 of *Lecture Notes in Control and Information Sciences (LNCS)*, pp. 152–160, Springer Berlin / Heidelberg, 2004.

[Dub07]   A. K. Dubey and V. Yadava. "Laser beam machining - A review". *International Journal of Machine Tools and Manufacture*, Vol. 48, No. 6, pp. 609–628, 2007.

[Dwo95]   R. Dworkowski and P. Wojcik. "Computer control for a high speed, precision laser cutting system". In: *Proceedings of the Canadian Conference on Electrical and Computer Engineering*, Vol. 1, pp. 187–189, 1995.

[Egg04]   G. Eggers, J. Klein, J. Blank, and S. Hassfeld. "Piezosurgery: an ultrasound device for cutting bone and its use and limitations in maxillofacial surgery.". *British Journal of Oral and Maxillofacial Surgery*, Vol. 42, No. 5, pp. 451–453, 2004.

[Eng03]   D. Engel. *Sensorgestützte Robotersteuerung für den Einsatz in der Chirurgie*. Doctoral thesis, Universität Karlsruhe (TH), Germany, 2003.

[Eyr05]   G. K. Eyrich. "Laser-osteotomy induced changes in bone". *Medical Laser Application*, Vol. 20, pp. 25–36, 2005.

[Fau06]   R. A. Faust, Ed. *Robotics in surgery: history, current, and future applications*. Nova Science Publishers, Inc., 2006.

[Fed03]   P. A. Federspil, B. Plinkert, and P. K. Plinkert. "Experimental robotic milling in skull-base surgery.". *Computer Aided Surgery*, Vol. 8, No. 1, pp. 42–48, 2003.

[Fir77]   A. Firester, M. Eller, and P. Sheng. "Knife-edge scanning measurements of subwavelength focused light beams". *Applied Optics*, Vol. 16, No. 7, pp. 1971–1974, 1977.

[Fit00]   J. M. Fitzpatrick and M. Sonka. *Handbook of Medical Imaging: Medical Image Processing and Analysis*. Vol. 2, SPIE Press, 2000.

[Fit09]   J. Fitzpatrick. "Fiducial registration error and target registration error are uncorrelated". In: *Proceedings of SPIE Medical Imaging: Visualization, Image-Guided Procedures, and Modeling*, Vol. 7261, p. 726102, 2009.

[Fit98]   J. Fitzpatrick, J. West, and C. Maurer. "Predicting error in rigid-body point-based registration". *IEEE Transactions on Medical Imaging*, Vol. 17, No. 5, pp. 694–702, 1998.

[For93]   M. Forrer, M. Frenz, V. Romano, H. Altermatt, H. Weber, A. Silenok, M. Istomyn, and V. Konov. "Bone-Ablation Mechanism Using CO_2 Lasers of Different Pulse Duration and Wavelength". *Applied Physics B Photophysics and Laser Chemistry*, Vol. B56, pp. 104–112, 1993.

[Fre03]   M. Frentzen, W. Götz, M. Ivanenko, S. Afilal, M. Werner, and P. Hering. "Osteotomy with 80-μs CO_2 laser pulses - histological results". *Lasers in Medical Science*, Vol. 18, No. 2, pp. 119–124, 2003.

Bibliography

[Fri03]   M. Fridenfalk and G. Bolmsjo. "Design and validation of a universal 6D seam tracking system in robotic welding based on laser scanning". *Industrial Robot: An International Journal*, Vol. 30, No. 5, pp. 437–448, 2003.

[Gil05]   A. Gillner, J. Holtkam, C. Hartmann, A. Olowinsky, J. Gedicke, K. Klages, L. Bosse, and A. Bayer. "Laser applications in microtechnology". *Journal of Materials Processing Technology*, Vol. 167, No. 2-3, pp. 494–498, 2005.

[Gol64a]   L. Goldman, P. Hornby, R. Meyer, and B. Goldman. "Impact of the Laser on Dental Caries". *Nature*, Vol. 203, p. 417, 1964.

[Gol64b]   L. Goldman, J. M. Igelman, and D. F. Richfield. "Impact of the Laser on Nevi And Melanomas". *Archives of Dermatology*, Vol. 90, pp. 71–75, 1964.

[Gol70]   L. Goldman, R. Rockwell, Z. Naprstek, V. Siler, R. Hoefer, C. Hobeika, K. Hishimoto, T. Polanyi, and H. Bredmeier. "Some Parameters of High Output CO_2 Laser Experimental Surgery". *Nature*, Vol. 228, pp. 1344–1345, 1970.

[Gra07]   M. de Graaf. *Sensor-guided Robotic Laser Welding*. Doctoral thesis, Universiteit Twente, Netherlands, 2007.

[Gru03]   M. Grupp, T. Seefeld, and F. Vollertsen. "Laser Beam Welding with Scanner". In: *Proceedings of the International WLT-Conference on Lasers in Manufacturing*, pp. 375–379, AT-Fachverlag, 2003.

[Hac97]   A. Hace, K. Jezernik, B. Curk, and M. Terbuc. "Robust motion control of XY table for laser cutting machine". In: *IEEE Industrial Electronics Society*, pp. 1097–1102, 1997.

[Hag08]   U. Hagn, M. Nickl, S. Jorg, G. Passig, T. Bahls, A. Nothhelfer, F. Hacker, L. Le-Tien, A. Albu-Schaffer, R. Konietschke, M. Grebenstein, R. Warpup, R. Haslinger, M. Frommberger, and G. Hirzinger. "The DLR MIRO: a versatile lightweight robot for surgical applications". *Industrial Robot: An International Journal*, Vol. 35, No. 4, pp. 324–336, 2008.

[Haj01]   J. Hajnal, D. Hawkes, and D. Hill. *Medical image registration*. CRC Press, 2001.

[Han99]   D. Hand, C. Peters, and J. Jones. "Online focus control sensor for laser welding offers improved reliability without process intrusion". *Sensor Review*, Vol. 19, No. 4, pp. 265–267, 1999.

[Hib97]   R. Hibst. *Technik, Wirkungsweise und medizinische Anwendung von Holmium- und Erbium-Lasern.* ecomed, 1997.

[Hil98]   J. W. Hills and J. F. Jensen. "Telepresence technology in medicine: principles and applications". *Proceedings of the IEEE*, Vol. 86, No. 3, pp. 569–580, 1998.

[Hoc07]   N. Hockstein, C. Gourin, R. Faust, and D. Terris. "A history of robots: from science fiction to surgical robotics". *Journal of Robotic Surgery*, Vol. 1, No. 2, pp. 113–118, July 2007.

[Hof08]   M. Hofer, E. Dittrich, C. Scholl, T. Neumuth, M. Strauss, A. Dietz, T. Lüth, and G. Strauss. "First clinical evaluation of the navigated controlled drill at the lateral skull base". In: *Medicine Meets Virtual Reality 16 - parallel, combinatorial, convergent: NextMed by Design*, Vol. 132 of *Studies in Health Technology and Informatics*, pp. 171–173, 2008.

[Hoh09]   B. Hohlweg-Majert, H. Deppe, M. Metzger, S. Stopp, K. Wolff, and T. Lueth. "Bone treatment laser-navigated surgery". *Lasers in Medical Science*, Vol. Online First, 2009.

[Hom09]   O. Homburg, B. Guetlich, F. Toennissen, T. Mitra, and L. Aschke. "ZnSe aspherical microlens systems enable new beam shaping approaches for CO_2-lasers". In: *SPIE 7196*, p. 71960B, 2009.

[Hor87]   B. K. P. Horn. "Closed-form solution of absolute orientation using unit quaternions". *JOSA A*, Vol. 4, pp. 629–642, 1987.

[Hui02]   J. Huisson. "Robotic laser welding: seam sensor and laser focal frame registration". *Robotica*, Vol. 20, pp. 261–268, 2002.

[Iar08]   G. Iaria. "Clinical, morphological, and ultrastructural aspects with the use of Er:YAG and Er,Cr:YSGG lasers in restorative dentistry". *General Dentistry*, Vol. 56, No. 7, pp. 636–639, 2008.

[Ion05]   J. Ion. *Laser processing of engineering materials: principles, procedure and industrial application.* Butterworth-Heinemann, 2005.

[Ise94]   H. Iseki, H. Kawamura, T. Tanikawa, H. Kawabatake, T. Taira, K. Takakura, T. Dohi, and N. Hata. "An image-guided stereotactic system for neurosurgical operations". *Stereotactic and Functional Neurosurgery*, Vol. 63, No. 1-4, pp. 130–138, 1994.

[Ish08]   I. Ishikawa, A. Aoki, and A. A. Takasaki. "Clinical application of erbium:YAG laser in periodontology". *Journal of the International Academy of Periodontology*, Vol. 10, No. 1, pp. 22–30, 2008.

[Iva00a]   M. Ivanenko, G. Eyrich, E. Bruder, and P. Hering. "In vitro incision of bone tissue with a Q-switch CO_2 laser. Histological examination.". *Lasers in the Life Sciences*, Vol. 9, pp. 171–179, 2000.

[Iva00b]   M. Ivanenko, T. Mitra, and P. Hering. "Hard tissue ablation with sub-μs CO_2 laser pulses with the use of an air-water spray". In: *Proceedings of SPIE Optical Biopsy and Tissue Optics*, 2000.

[Iva02]   M. Ivanenko, S. Fahimi-Weber, T. Mitra, W. Wierich, and P. Hering. "Bone Tissue Ablation with sub-μs Pulses of a Q-switch CO_2 Laser: Histological Examination of Thermal Side-Effects". *Lasers in Medical Sciences*, Vol. 17, No. 6, pp. 258–264, 2002.

[Iva05a]   M. Ivanenko, R. Sader, S. Afilal, M. Werner, M. Hartstock, C. von Hänisch, S. Milz, W. Erhardt, H.-F. Zeilhofer, and P. Hering. "In vivo animal trials with a scanning CO_2 laser osteotome.". *Lasers in Surgery and Medicine*, Vol. 37, No. 2, pp. 144–148, 2005.

[Iva05b]   M. Ivanenko, M. Werner, S. Afilal, M. Klasing, and P. Hering. "Ablation of hard bone tissue with pulsed CO_2 lasers". *Medical Laser Applications*, Vol. 20, pp. 13–23, 2005.

[Iva06]   M. Ivanenko, M. Werner, M. Klasing, and P. Hering. "System development and clinical studies with a scanning CO_2 laser osteotome". In: *Proceedings of the SPIE Optical Interactions with Tissue and Cells XVII*, Vol. 6084, pp. 109–114, 2006.

[Iva98a]   M. Ivanenko and P. Hering. "Wet bone ablation with mechanically Q-switched high-repetition-rate CO_2 laser". *Applied Physics B: Lasers and Optics*, Vol. 67, No. 3, pp. 395–397, 1998.

[Iva98b]   M. Ivanenko and P. Hering. "Hard tissue ablation with a mechanically Q-switched CO_2 laser". In: *Proceedings of SPIE Thermal Therapy, Laser Welding and Tissue Interactions*, Vol. 3565, pp. 110–115, 1998.

[Jac93]   S. L. Jacques. "Role of tissue optics and pulse duration on tissue effects during high-power laser irradiation". *Applied Optics*, Vol. 32, No. 13, pp. 2447–2454, 1993.

[Jam86]   H. James, C. Wiley, and S. Schneider. "The effect of carbon dioxide laser irradiation on cranial bone healing - An experimental study". *Childs Nervous System*, Vol. 2, pp. 248–251, 1986.

[Jen09]   Jenoptik Automatisierungstechnik GmbH. online: `http://www.automation-jenoptik.com`, Oct. 2009.

[Jim07]   C. Jimin, Y. Jianhua, Z. Shuai, Z. Tiechuan, and G. Dixin. "Parameter optimization of non-vertical laser cutting". *The International Journal of Advanced Manufacturing Technology*, Vol. 33, No. 5, pp. 469–473, 2007.

[Joh98]   T. F. Johnston. "Beam Propagation (M^2) measurement made as easy as it gets: the four-cuts method". *Applied Optics*, Vol. 37, No. 21, pp. 4840–4850, 1998.

[Jos01]   L. Joskowicz and R. H. Taylor. "Computers in imaging and guided surgery". *Computing in Science & Engineering*, Vol. 3, No. 5, pp. 65–72, 2001.

[Kah07]   L. A. Kahrs, M. Werner, F. Knapp, S. Lu, J. Raczkowsky, J. Schipper, M. Ivanenko, H. Wörn, P. Hering, and T. Klenzner. "Video Camera Based Navigation of a Laser Beam for Micro Surgery Bone Ablation at the Skull Base Setup and Initial Experiments". In: *Advances in Medical Engineering, Springer Proceedings in Physics*, Vol. 114, pp. 219–223, 2007.

[Kah08]   L. A. Kahrs, J. Raczkowsky, M. Werner, F. Knapp, M. Mehrwald, P. Hering, J. Schipper, T. Klenzner, and

H. Wörn. "Visual servoing of laser ablation based cochleostomy". In: *Proceedings SPIE Medical Imaging: Visualization, Image-Guided Procedures, and Modeling*, Vol. 6918, pp. 69182C–11, 2008.

[Kah09a]   L. A. Kahrs. *Bildverarbeitungsunterstützte Laserknochenablation am humanen Felsenbein*. Doctoral thesis, Universität Karlsruhe (TH), Germany, 2009.

[Kah09b]   L. A. Kahrs, J. Burgner, T. Klenzner, J. Raczkowsky, J. Schipper, and H. Wörn. "Planning and simulation of microsurgical laser bone ablation". *International Journal of Computer Assisted Radiology and Surgery*, Vol. 5, No. 2, pp. 155–162, 2009.

[Kan08]   H. Kang, J. Oh, and A. Welch. "Investigations on laser hard tissue ablation under various environments". *Physics in Medicine and Biology*, Vol. 53, No. 12, pp. 3381–3390, 2008.

[Kar72]   R. Karp. "Reducibility Among Combinatorial Problems". *Complexity of Computer Computations*, pp. 85–103, 1972.

[Keu07]   J. Keuster, J. R. Duflou, and J. P. Kruth. "Monitoring of high-power CO2 laser cutting by means of an acoustic microphone and photodiodes". *The International Journal Of Advanced Manufacturing Technology*, Vol. 35, No. 1, pp. 115–126, 2007.

[Kha00]   R. Khadem, C. C. Yeh, M. Sadeghi-Tehrani, M. R. Bax, J. A. Johnson, J. N. Welch, E. P. Wilkinson, and R. Shahidi. "Comparative tracking error analysis of five different optical tracking systems". *Computer Aided Surgery*, Vol. 5, No. 2, pp. 98–107, 2000.

[Kim04]   D. Kim, D. Kim, H. Owada, N. Hata, and T. Dohi. "An Er:YAG Laser Bone Cutting Manipulator for Precise Rotational Acetabular Osteotomy". In: *Proceedings of the IEEE Engineering in Medicine and Biology (EMB)*, Vol. 4, pp. 2750–2753, 2004.

[Kla08]   M. Klasing. *Konstruktion und Evaluierung eines CO_2-Laserosteotoms*. Master-Thesis, University of Applied Sciences Koblenz, RheinAhrCampus Remagen, 2008.

[Kla83]   Y. Klang and R. Lang. "Measuring focused Gaussian beam spot sizes: a practical method". *Applied Optics*, Vol. 22, No. 9, pp. 1296–1297, 1983.

[Kle09]   T. Klenzner, C. C. Ngan, F. B. Knapp, H. Knoop, J. Kromeier, A. Aschendorff, E. Papastathopoulos, J. Raczkowsky, H. Wörn, and J. Schipper. "New strategies for high precision surgery of the temporal bone using a robotic approach for cochlear implantation.". *European Archives of Oto-Rhino-Laryngology*, Vol. 266, No. 7, pp. 955–960, 2009.

[Kno07]   H. Knoop, J. Raczkowsky, U. Wyslucha, T. Fiegele, G. Eggers, and H. Wörn. "Integration of intraoperative imaging and surgical robotics to increase their acceptance". *International Journal of Computer Assisted Radiology and Surgery*, Vol. 1, No. 5, pp. 243–251, 2007.

[Koc02]   A. Kochan. "A time of change for welding". *Assembly Automation*, Vol. 22, No. 1, pp. 29–35, 2002.

[Kor03]   W. Korb, D. Engel, R. Boesecke, G. Eggers, B. Kotrikova, R. Marmulla, J. Raczkowsky, H. Wörn, J. Mühling, and S. Hassfeld. "Development and first patient trial of a surgical robot for complex trajectory milling". *Computer Aided Surgery*, Vol. 8, No. 5, pp. 247–256, 2003.

[Kor07]   J. Kordt. *Konturnahes Laserstrahlstrukturieren für Kunststoffspritzgießwerkzeuge*. Doctoral thesis, RWTH Aachen, Germany, 2007.

[Kuh05]   C. Kuhlmann, F. Pude, C. Bishup, S. Krömer, L. Kirsch, A. Andreae, K. Wacker, and S. Schmolke. "Evaluation of potential risks of abrasive water jet osteotomy in-vivo". *Biomedizinische Technik*, Vol. 50, No. 10, pp. 337–342, 2005.

[KUK07]   KUKA Roboter GmbH. "Instruction Manual". 2007.

[Kut08]   J. J. Kuttenberger, S. Stübinger, A. Waibel, M. Werner, M. Klasing, M. Ivanenko, P. Hering, B. V. Rechenberg, R. Sader, and H.-F. Zeilhofer. "Computer-Guided CO_2-Laser Osteotomy of the Sheep Tibia: Technical Prerequisites and First Results.". *Photomedicine and Laser Surgery*, Vol. 26, No. 2, pp. 129–136, 2008.

Bibliography

[Kut10]   J. J. Kuttenberger, A. Waibel, S. Stübinger, M. Werner, M. Klasing, M. Ivanenko, P. Hering, B. von Rechenberg, R. Sader, and H.-F. Zeilhofer. "Bone healing of the sheep tibia shaft after carbon dioxide laser osteotomy: histological results.". *Lasers Med Sci*, Vol. 25, No. 2, pp. 239–249, Mar 2010.

[Kwo88]   Y. Kwoh, J. Hou, E. Jonckheere, and S. Hayati. "A robot with improved absolute positioning accuracy for CT guided stereotactic brain surgery". *IEEE Transactions on Biomedical Engineering*, Vol. 35, No. 2, pp. 153–160, 1988.

[Lan02]   B. I. Lange, T. Brendel, and G. Hüttmann. "Temperature Dependence of Light Absorption in Water at Holmium and Thulium Laser Wavelengths". *Applied Optics*, Vol. 41, No. 27, pp. 5797–5803, 2002.

[Las09]   Laser Mech Europe. online: http://www.lasermech.com, Oct. 2009.

[LaV06]   S. LaValle. *Planning Algorithms*. Cambridge University Press, 2006.

[Lit64]   N. H. Lithwick, M. K. Healy, and J. Cohen. "Micro-Analysis of bone by laser microprobe.". *Surgical Forum*, Vol. 15, pp. 439–441, 1964.

[Lor87]   W. Lorensen and H. Cline. "Marching cubes: A high resolution 3d surface construction algorithm". *ACM Computer Graphics*, Vol. 21, pp. 163–169, 1987.

[Mad02]   O. Madsen, C. B. Sorensen, R. Larsen, L. Overgaard, and N. J. Jacobsen. "A system for complex robotic welding". *Industrial Robot: An International Journal*, Vol. 29, No. 2, pp. 127–131, 2002.

[Mai60]   T. H. Maiman. "Stimulated optical radiation in ruby". *Nature*, Vol. 187, pp. 493–494, 1960.

[Mai98]   J. Maintz and M. Viergever. "A survey of medical image registration". *Medical Image Analysis*, Vol. 2, No. 1, pp. 1–36, 1998.

[Maj03]   J. D. Majumdar and I. Manna. "Laser Material Processing". In: *Proceedings in Engineering Sciences*, Vol. 28, pp. 495–562, Sadhana-Academy, 2003.

[Maj09]   O. Majdani, T. Rau, S. Baron, H. Eilers, C. Baier, B. Heimann, T. Ortmaier, S. Bartling, T. Lenarz, and M. Leinung. "A robot-guided minimally invasive approach for cochlear implant surgery: preliminary results of a temporal bone study". *International Journal of Computer Assisted Radiology and Surgery*, Vol. 4, No. 5, pp. 475–486, 2009.

[Mar04]   G. Marshall. *Handbook of optical and laser scanning*. CRC Press, 2004.

[Mas05]   C. R. Mascott. "Comparison of Magnetic Tracking and Optical Tracking by Simultaneous Use of Two Independent Frameless Stereotactic Systems". *Neurosurgery*, Vol. 57, No. 4, pp. 295–301, 2005.

[McK86]   A. McKenzie. "A three-zone model of soft-tissue damage by CO_2 laser". *Physics in Medicine and Biology*, Vol. 31, No. 9, pp. 967–983, 1986.

[McK90]   A. McKenzie. "Physics of thermal processes in laser-tissue interaction". *Physics in Medicine and Biology*, Vol. 35, No. 9, pp. 1175–1209, 1990.

[MED09a]   MED-EL Medical Electronics. online: `http://www.medel.com`, Oct. 2009.

[Med09b]   Medical Product Safety Network (MedSun). online: `http://www.fda.gov/cdrh/medsun`, Apr. 2009.

[Meh10]   M. Mehrwald, J. Burgner, C. Platzek, C. Feldmann, J. Raczkowsky, and H. Wörn. "Analysis of the short-pulsed CO_2 laser ablation process for optimising the processing performance for cutting bony tissue". In: *Proceedings SPIE Photonics West BIOS - Optical Interactions with Tissues and Cells XXI*, Vol. 7562, 2010.

[Mei04]   J. Meijer. "Laser beam machining (LBM), state of the art and new opportunities". *Journal of Materials Processing Technology*, Vol. 149, No. 1-3, pp. 2–17, 2004.

[Mei98]   D. Meister, P. Pokrandt, and A. Both. "Milling accuracy in robot assisted orthopaedic surgery". In: *Proc. 24th Annual Conference of the IEEE Industrial Electronics Society IECON '98*, Vol. 4, pp. 2502–2505, 1998.

Bibliography

[Mil09]   F. Milani. *Leistungs- und Geometriekorrelation für die Scan-nergesteuerte Laserknochenablation unter Verwendung eines Spiegelgelenkarms.* Diploma thesis, Universität Karlsruhe (TH), Germany, 2009.

[Mir08]   M. Mir, J. Meister, R. Franzen, S. Sabounchi, F. Lampert, and N. Gutknecht. "Influence of water-layer thickness on Er:YAG laser ablation of enamel of bovine anterior teeth". *Laser in Medical Science*, Vol. 23, No. 4, pp. 451–457, 2008.

[Moe09]   H. Moennich, J. Raczkowsky, D. Stein, and H. Wörn. "Increasing the accuracy with a rich sensor system for robotic laser osteotomy". In: *IEEE Sensors*, pp. 1684 –1688, 2009.

[Mül09]   M. Müller. *Transferring cutting trajectories into pulse sequences and optimized robot locations for laser osteotomy.* Diploma thesis, Universität Karlsruhe (TH), 2009.

[Nie07]   M. H. Niemz. *Laser-Tissue Interactions - Fundamentals and Applications.* Springer Berlin, 3rd Ed., 2007.

[Ore09]   Oregon Medical Laser Center. online: `http://omlc.ogi.edu/spectra/water/abs/index.html`, Sep. 2009.

[Oro09]   Orocos Project. online: `http://www.orocos.org/`, Sep. 2009.

[Pap07]   M. Papadaki, A. Doukas, W. Farinelli, L. Kaban, and M. Troulis. "Vertical ramus osteotomy with Er:YAG laser: a feasibility study". *International Journal of Oral and Maxillofacial Surgery*, Vol. 36, No. 12, pp. 1193–1197, 2007.

[Pas04]   A. Pashkevich, A. Dolgui, and O. Chumakov. "Multiobjective optimization of robot motion for laser cutting applications". *International Journal of Computer Integrated Manufacturing*, Vol. 17, No. 2, pp. 171–183, 2004.

[Pec09]   I. Pechlivanis, G. Kiriyanthan, M. Engelhardt, M. Scholz, S. Lücke, A. Harders, and K. Schmieder. "Percutaneous placement of pedicle screws in the lumbar spine using a bone mounted miniature robotic system: first experiences and accuracy of screw placement.". *Spine*, Vol. 34, No. 4, pp. 392–398, 2009.

[Pet05]   H. Peters, H. Knoop, W. Korb, S. Ghanai, J. Raczkowsky, M. Werner, M. Klasing, M. Ivanenko, S. Hassfeld, P. Hering, and H. Wörn. "Bringing Laser for Osteotomy into the Operation Theatre". In: *Proceedings of the International Conference and Exhibiton Computer Assisted Radiology and Surgery (CARS)*, p. 1364, 2005.

[Pir02]   J. Pires, A. Loureiro, T. Godinho, P. Ferreira, B. Fernando, and J. Morgado. "Object oriented and distributed software applied to industrial robotic welding". *Industrial Robot: An International Journal*, Vol. 29, No. 2, pp. 149–161, 2002.

[Pla09]   C. Platzek. *Modellierung und Evaluierung von Schnittechniken für die roboterassistierte Laserosteotomie.* Diploma thesis, Universität Karlsruhe (TH), Germany, 2009.

[Pow98]   J. Powell. *CO_2 Laser Cutting.* Springer, 2nd Ed., 1998.

[Pre07]   B. Preim and D. Bartz. *Visualization in Medicine: Theory, Algorithms, and Applications. Computer Graphics*, Morgan Kaufmann Publishers Inc. 2007.

[Rei08]   G. Reinhart, U. Munzert, and W. Vogl. "A programming system for robot-based remote-laser-welding with conventional optics". *CIRP Annals - Manufacturing Technology*, Vol. 57, No. 1, pp. 37–40, 2008.

[Riz05]   P. Rizun and G. Sutherland. "Surgical Laser Augmented with Haptic Feedback and Visible Trajectory". In: *IEEE Virtual Reality*, pp. 257–260, 2005.

[Roc09]   Rockwell Laser Industries. "Laser Accident Database". online: `http://www.rli.com/resources/accident.aspx`, Apr. 2009.

[Rog01]   G. Roggero, F. Scotti, F. Sessa, and V. Piuri. "Quality analysis measurement for laser cutting". In: *Proceedings of the IEEE International Workshop on Virtual and Intelligent Measurement Systems*, pp. 41–44, 2001.

[Rom09]   U. Romeo, A. Del Vecchio, G. Palaia, G. Tenore, P. Visca, and C. Maggiore. "Bone damage induced by different cutting instruments–an in vitro study.". *Brazilian Dental Journal*, Vol. 20, No. 2, pp. 162–168, 2009.

[Roo00]   B. Rooks. "Lasers become the acceptable face of precision welding and cutting". *Industrial Robot: An International Journal*, Vol. 27, No. 2, pp. 103–107, 2000.

[Rup03]   S. Rupprecht, K. Tangermann, P. Kessler, F. Neukam, and J. Wiltfang. "Er:YAG laser osteotomy directed by sensor controlled systems". *Journal of Cranio-Maxillofacial Surgery*, Vol. 31, pp. 337–342, 2003.

[Rup04]   S. Rupprecht, K. Tangermann-Gerk, J. Wiltfang, F. Neukam, and A. Schlegel. "Sensor-based laser ablation for tissue specific cutting: an experimental study". *Lasers in Medical Science*, Vol. 19, No. 2, pp. 81–88, 2004.

[Sal08]   A. Salami, M. Dellepiane, F. Mora, B. Crippa, and R. Mora. "Piezosurgery in the cochleostomy through multiple middle ear approaches.". *International Journal of Pediatric Otorhinolaryngology*, Vol. 72, No. 5, pp. 653–657, 2008.

[Sch07]   A. P. Schulz, K. Seide, C. Queitsch, A. von Haugwitz, J. Meiners, B. Kienast, M. Tarabolsi, M. Kammal, and C. Jürgens. "Results of total hip replacement using the Robodoc surgical assistant system: clinical outcome and evaluation of complications for 97 procedures.". *The International Journal of Medical Robotics and Computer Assisted Surgery*, Vol. 3, No. 4, pp. 301–306, 2007.

[Sch08]   F. Schmitt, M. Funck, A. Boglea, and R. Poprawe. "Development and application of miniaturized scanners for laser beam micro-welding". *Microsystem Technologies*, Vol. 14, No. 12, pp. 1861–1869, 2008.

[Sch81]   M. Schneider and W. Webb. "Measurement of submicron laser beam radii". *Applied Optics*, Vol. 20, No. 8, pp. 1382–1388, 1981.

[Sha09]   R. Shamir and L. Joskowicz. "Worst-case analysis of target localization errors in fiducial based rigid body registration". In: *Proceedings SPIE*, Vol. 7259, p. 725938, 2009.

[Sho03]   M. Shoham, M. Burman, E. Zehavi, L. Joskowicz, E. Batkilin, and Y. Kunicher. "Bone-mounted miniature robot for surgical procedures: Concept and clinical applications". *IEEE Transactions on Robotics and Automation*, Vol. 19, No. 5, pp. 893–901, 2003.

[Sic08]   B. Siciliano and O. Khatib, Eds. *Springer Handbook of Robotics*. Springer, Berlin, Heidelberg, 2008.

[Sil04]   W. Silfvast. *Laser Fundamentals*. Cambridge University Press, 2nd Ed., 2004.

[Smi06]   J. Smith. "Robots and lasers: the future of cardiac tissue ablation?". *The International Journal of Medical Robotics and Computer Assisted Surgery*, Vol. 2, No. 4, pp. 329–332, 2006.

[Sol07]   C. Solares and M. Strome. "Transoral Robot-Assisted CO_2 Laser Supraglottic Laryngectomy: Experimental and Clinical Data". *The Laryngoscope*, Vol. 117, No. 5, pp. 817–820, 2007.

[StO02]   L. St-Onge. "A mathematical framework for modeling the compositional depth profiles obtained by pulsed laser ablation". *Journal of Analytical Atomic Spectrometry*, Vol. 17, No. 9, pp. 1083–1089, 2002.

[Stä04]   Stäubli. "Arm – RX90 series 90B family Characteristics". 2004.

[Ste03a]   W. Steen. *Laser Material Processing*. Springer London, 3rd Ed., 2003.

[Ste03b]   W. Steen. "Laser material processing - an overview". *Journal of Optics A: Pure and Applied Optics*, Vol. 5, pp. 3–7, 2003.

[Ste06]   J. Stemmann and R. Zunke. "Robot Task Planning for Laser Remote Welding". In: *Proceedings of Operations Research*, Vol. Volume 2005 of *Operations Research Proceedings*, Springer Berlin Heidelberg, 2006.

[Sto05]   A. Stojanovic and D. Suput. "Strategic Planning in Topography-guided Ablation of Irregular Astimatism After Laser Refractive Surgery". *Journal of Refractive Surgery*, Vol. 21, pp. 369–376, 2005.

[Sto07a]   P. J. Stolka, M. Waringo, D. Henrich, S. H. Tretbar, and P. A. Federspil. "Robot-Based 3D Ultrasound Scanning and Registration with Infrared Navigation Support". In: *Proceedings of the IEEE International Conference on Robotics and Automation (ICRA)*, pp. 2635–2641, 2007.

[Sto07b]   S. Stopp, D. Svejdar, H. Deppe, and T. C. Lueth. "A new method for optimized laser treatment by laser focus navigation and distance visualization.". In: *Proceedings of the IEEE Engineering in Medicine and Biology (EMB)*, pp. 1738–1741, 2007.

[Sto08a]   S. Stopp. *Ein integriertes System für die Mund-, Kiefer- und Gesichtschirurgie und Implantologie. Fortschritt-Berichte*, VDI-Verlag, 2008.

[Sto08b]   S. Stopp, H. Deppe, and T. Lueth. "A new concept for navigated laser surgery.". *Lasers in Medical Science*, Vol. 23, No. 3, pp. 261–266, 2008.

[Sto08c]   S. Stopp, D. Svejdar, E. von Kienlin, H. Deppe, and T. C. Lueth. "A new approach for creating defined geometries by navigated laser ablation based on volumetric 3-D data.". *IEEE Transactions on Biomedical Engineering*, Vol. 55, No. 7, pp. 1872–1880, 2008.

[Stü07a]   S. Stübinger, C. Kober, H.-F. Zeilhofer, and R. Sader. "Er:YAG laser osteotomy based on refined computer-assisted presurgical planning: first clinical experience in oral surgery". *Photomedicine and Laser Surgery*, Vol. 25, No. 1, pp. 3–7, 2007.

[Stü07b]   S. Stübinger, C. Landes, O. Seitz, and R. Sader. "Er:YAG Laser Osteotomy for Intraoral Bone Grafting Procedures: A Case Series With a Fiber-Optic Delivery System". *Journal of Periodontology*, Vol. 78, No. 12, pp. 2389–2394, 2007.

[Stü09]   S. Stübinger, S. Ghanaati, B. Saldamli, C. Kirkpatrick, and R. Sader. "Er:YAG Laser Osteotomy: Preliminary Clinical and Histological Results of a New Technique for Contact-Free Bone Surgery". *European Surgical Research*, Vol. 42, pp. 150–156, 2009.

[Szy03]   D. Szymanski and T. Lüth. "Navigated control in dental implantology". In: *Proceedings of the 17th International Congress and Exhibition Computer Assisted Radiology and Surgery (CARS)*, p. 1409, 2003.

[Tan05]   H. Tang, H. Van Brussel, J. Vander Sloten, D. Reynaerts, and P. Koninckx. "Implementation of an Intuitive Writing Interface and a Laparoscopic Robot for Gynaecological Laser

Assisted Surgery". *Proceedings of the Institution of Mechanical Engineers, Part H: Journal of Engineering in Medicine,* Vol. 213, No. 4, pp. 239–302, 2005.

[Tay03a] R. H. Taylor and D. Stoianovici. "Medical robotics in computer-integrated surgery". *IEEE Transactions on Robotics and Automation,* Vol. 19, No. 5, pp. 765–781, 2003.

[Tay03b] R. Taylor and L. Joskowicz. *Computer-integrated Surgery and Medical Robotics,* Chap. 29, pp. 325–353. McGraw-Hill, 2003.

[Tay06] R. H. Taylor. "A Perspective on Medical Robotics". *Proceedings of the IEEE,* Vol. 94, No. 9, pp. 1652–1664, 2006.

[Tay94a] M. Tayal, V. Barnekov, and K. Mukherjee. "Focal point location in laser meachining of thick hard wood". *Journal of Materials Science Letters,* Vol. 13, pp. 644–646, 1994.

[Tay94b] R. H. Taylor, B. D. Mittelstadt, H. A. Paul, W. Hanson, P. Kazanzides, J. F. Zuhars, B. Williamson, B. L. Musits, E. Glassman, and W. L. Bargar. "An image-directed robotic system for precise orthopaedic surgery". *IEEE Transactions on Robotics and Automation,* Vol. 10, No. 3, pp. 261–275, 1994.

[Tso06] G. Tsoukantas and G. Chryssolouris. "Theoretical and experimental analysis of the remote welding process on thin, lap-joined AISI 304 sheets". *The International Journal of Advanced Manufacturing Technology,* Vol. 35, No. 9-10, pp. 880–894, 2006.

[Tso08] G. Tsoukantas, A. Stournaras, and G. Chryssolouris. "An experimental investigation of remote welding with CO2 and Nd: YAG laser-based systems". *Journal of Laser Applications,* Vol. 20, No. 1, pp. 50–58, 2008.

[Ven94] V. Venugopalan, N. S. Nishioka, and B. B. Mikic. "The Effect of Laser Parameters on the Zone of Thermal Injury Produced by Laser Ablation of Biological Tissue". *Journal of Biomechanical Engineering,* Vol. 116, No. 1, pp. 62–70, 1994.

[Ver04] T. Vercellotti. "Technological characteristics and clinical indications of piezoelectric bone surgery". *Minerva Stomatologica,* Vol. 53, No. 5, pp. 207–214, 2004.

Bibliography

[Vog03]   A. Vogel and V. Venugopalan. "Mechanisms of Pulsed Laser Ablation of Biological Tissues". *Chemical Reviews*, Vol. 103, No. 2, pp. 577–644, 2003.

[Wai07]   R. Waiboer. *Dynamic Modelling, Identification and Simulation of Industrial Robots - for Off-line Programming of Robotised Laser Welding*. Doctoral thesis, Universiteit Twente, Netherlands, 2007.

[Wan00]   R. K. Wang. "Modelling optical properties of soft tissue by fractal distribution of scatterers". *Journal of Modern Optics*, Vol. 47, No. 1, pp. 103 – 120, 2000.

[Wan06]   Y. Wang, S. E. Butner, and A. Darzi. "The Developing Market for Medical Robotics". *Proceedings of the IEEE*, Vol. 94, No. 9, pp. 1763–1771, 2006.

[Wan99]   Y. Wang, Q. Bao, S. Wang, and L. Wu. "Laser Garment Cut Robot System: Design and Realization". In: *Proceedings of the IEEE International Conference on Robotics and Automation (ICRA)*, pp. 1084–1089, 1999.

[Web04]   C. Webb and J. Jones. *Handbook of laser technology and applications*. Taylor & Francis, 2004.

[Wel95]   A. J. Welch and M. J. C. van Gemert. *Optical-thermal response of laser-irradiated tissue*. Springer, 1995.

[Wer06]   M. Werner. *Ablation of hard biological tissue and osteotomy with pulsed CO_2 lasers*. Doctoral thesis, Heinrich-Heine-Universität Düsseldorf, Germany, 2006.

[Wer07a]   M. Werner, M. Ivanenko, D. Harbecke, M. Klasing, H. Steigerwald, and P. Hering. "CO_2 laser milling of hard tissue". In: *Proceedings of the SPIE Optical Interactions with Tissue and Cells XVIII*, Vol. 6435, 2007.

[Wer07b]   M. Werner, M. Klasing, M. Ivanenko, D. Harbecke, H. Steigerwald, and P. Hering. "CO_2 laser free-form processing of hard tissue". In: *Proceedings of the SPIE Therapeutic Laser Applications and Laser-Tissue Interactions III*, Vol. 6632, 2007.

[Wid07]   G. Widmann. "Image-guided Surgery and Medical Robotics in the Cranial Area". *Biomedical Imaging and Intervention Journal*, Vol. 3, No. 1, p. e11, 2007.

[Wim09]   K. Wimmer. *Risk analysis of an Experimental Setup for Robot Assisted Laser Osteotomy.* Diploma thesis, Universität Karlsruhe (TH), Germany, 2009.

[Wör05]   H. Wörn, H. Peters, M. Ivanenko, and P. Hering. "LASER Based Osteotomy with Surgical Robots". In: *Biomedizinische Technik*, Vol. 50, pp. 25–26, 2005.

[Wör06]   H. Wörn. "Computer- and robot-aided head surgery". *Medical Technologies in Neurosurgery*, Vol. 98, pp. 51–61, 2006.

[Woo05]   B. J. Wood, H. Zhang, A. Durrani, N. Glossop, S. Ranjan, D. Lindisch, E. Levy, F. Banovac, J. Borgert, S. Krueger, J. Kruecker, A. Viswanathan, and K. Cleary. "Navigation with electromagnetic tracking for interventional radiology procedures: a feasibility study.". *Journal of Vascular and Interventional Radiology*, Vol. 16, No. 4, pp. 493–505, 2005.

[Xie05]   J. Xien, S. Huang, Z. Duan, Y. Shi, and S. Wen. "Correction of the image distortion for laser galvanometric scanning system". *Optics & Laser Technology*, Vol. 37, No. 4, pp. 305–311, 2005.

[Xu04]   D. Xu, M. Tan, X. Zhao, and Z. Tu. "Seam Tracking and Visual Control for Robotic Arc Welding Based on Structured Light Stereovision". *Internation Journal of Automation and Computing*, Vol. 1, pp. 63–75, 2004.

[Yan06]   Z. Yaniv and K. Cleary. "Image-Guided Procedures: A Review". Tech. Rep., Georgetown University Medical Center, 2006.

[Yan09]   Z. Yaniv, E. Wilson, D. Lindisch, and K. Cleary. "Electromagnetic tracking in the clinical environment.". *Medical Physics*, Vol. 36, No. 3, pp. 876–892, 2009.

[Yao05]   Y. Yao, H. Chen, and W. Zhang. "Time scale effects in laser material removal: a review". *The International Journal of Advanced Manufacturing Technology*, Vol. 26, No. 4, pp. 598–608, 2005.

[Yil97a]   B. Yilbas. "A study into CO_2 laser cutting process". *Heat and Mass Transfer*, Vol. 32, No. 3, pp. 175–180, 1997.

[Yil97b]   B. Yilbas and A. Kar. "Laser Spot Welding and Efficiency Consideration". *Journal of Materials Engineering and Performance*, Vol. 6, No. 6, pp. 766–770, 1997.

[Yil98]   B. S. Yilbas. "3-dimensional laser heating model including a moving heat source consideration and phase change process". *Heat and Mass Transfer*, Vol. 33, pp. 495–505, 1998.

[Zar61]   M. M. Zaret, G. M. Breinin, H. Schmidt, H. Ripps, I. M. Siegel, and L. R. Solon. "Ocular lesions produced by an optical maser (laser).". *Science*, Vol. 134, pp. 1525–1526, 1961.

[Zha07]   W. Zhang, Q. Chen, G. Zhang, Z. Sun, and D. Du. "Seam Tracking of Articulated Robot for Laser Welding Bbased on Visual Feedback Control". In: *Robotic Welding, Intelligence and Automation*, Vol. 362 of *Lecture Notes in Control and Information Sciences*, pp. 281–287, Springer Berlin / Heidelberg, 2007.

[Zha09]   Y. Zhang. *Optimization of Pulse Distribution for Short-Pulsed CO_2 Laser Ablation of Bone Tissue Based on Thermal Relaxation.* Diploma thesis, Karlsruhe Institute of Technology (KIT), Germany, 2010.

[Zwe87]   A. Zweig and H. Weber. "Mechanical and Thermal Parameters in Pulsed Laser Cutting of Tissue". *IEEE Journal of Quantum Electronics*, Vol. 23, No. 10, pp. 1787–1793, 1987.

[Zwe88]   A. D. Zweig, M. Frenz, V. Romano, and H. P. Weber. "A comparative study of laser tissue interaction at $2.94\mu m$ and $10.6\mu m$". *Applied Physics B: Lasers and Optics*, Vol. 47, No. 3, pp. 259–265, 1988.